ARTIST'S BOOK YEARBOOK 1994 - 95

ARTIST'S BOOK YEARBOOK 1994 - 95
Published by
MAGPIE PRESS
1. Hermitage Cottage,
Clamp Hill,
Stanmore,
Middx HA7 3JW
Spring '95

British Library Cataloguing in Publication Data
Artist's Book Yearbook. - 1994 - 95
I. Peixoto, Tanya.
686

ISBN 0-9523880-0-6

Printed in Great Britain by The Book Factory
35 / 37 Queensland Road, London N7 7AH

Cover design and Yearbook logo by Pip Thompson
Book design by Tanya Peixoto, John Bently
and Yolanda Christian (Design / DTP Tel. 0181 558 2039)

EDITORS

Tanya Peixoto, John Bently,
Stephanie Brown and Stefan Szczelkun

CONTRIBUTORS

Marshall Anderson, Natalie d'Arbeloff, John Bently, Ian Breakwell, Stephanie Brown, Johanna Drucker, Richard Essex, Simon Ford, Sibylle Fraser, Mike Horl, Sarah Jacobs, D. Jarvis, Roy Marchant, Stephen Mepsted, Louise Neaderland, John Nicholson, Isabella Oulton, Mark Pawson, Clive Phillpot, Martin Rogers, Stefan Szczelkun, Nikonova Tarshis, Matthew Tyson.

Artist's pages (special edition Yearbook): by Natalie d'Arbeloff, David Barton, John Bently, Les Bicknell, Braincell, Bob Cobbing, John Dilnot, The Dinosaur Press, Mikey Georgeson, Susan Johanknecht & Jenifer Newson, Tony Kemplen, Andrea Hill, Sarah Jacobs, Liliane Lijn, Andi McGarry, Brigid McLeer and Tone Myskja with Fine Art students from Univ. of Ulster, Stephen Mepsted, Jürgen Olbrich, Ti Parks, Simon Redington, Dawn Redwood, Jim Roberts.

ACKNOWLEDGEMENTS

Thank you to the American Printing History Association for permission to reprint "Underground Printing in Europe", by Sibylle Fraser, originally published in *Printing History* 16 (1986), the Journal of the American Printing History Association.

Special thanks to all the contributors, Pip Thompson for advice and support, to Wendy Reith, Stephen Mepsted andYolanda Christian for coming to my rescue

CONTENTS :

INTRODUCTION

Stephanie Brown's idea for an Artist's Book journal proposed at the first Bookmaker's Conference (Sept '93) was the starting point for this publication. Silvie Turner and Stefan Szczelkun guessed that the Arts Council might back a yearbook rather than a magazine and over coffee one day John Bently, Stefan and I put together an application which resulted in a small grant.

A year and seven months later, I am older and wiser, I have learnt how to hold editorial meetings in a pub and be sober enough to remind everyone the following day of the jobs they promised to do. I've learnt that money is very important but that working for nothing means you only have yourself to answer to. I've also met some brilliant people who have continually reassured me that it's going to be alright, (and believed them). The first Artist's Book Yearbook is born and with 23 contributors, as well as 22 pages by artists in the special edition, and a list of over 200 books produced this year, a lot of people have shown goodwill and enthusiasm. The emphasis has been on ideas rather than the practice of individuals and we have tried to pick up on odd and sometimes quirky areas of thought as well as the more general trends.

One of the limitations of the Yearbook, has to be the focus on London-based artists, with a few exceptions. I hope that this will not deter people from sending in contributions for consideration to next year's publication. Most people will understand that a book like this begins with who you know and who responds, but our aim is that it should grow beyond the network established in this first issue.

Meetin Judith Hoffberg (USA), editor of Umbrella Magazine and founder of ARLIS at the Blue Posts Pub (Sept '94), was a high point. As she pointed out, there is an Artist's Book scene in the UK, which is alive and demanding attention: Workshops in schools, bookart in colleges, collections in libraries, reviews in art magazines, launches and exhibitions, fairs, European connections as well as more contact with what's happening in the States via Umbrella and now JAB (Journal of Artist's Books) should encourage bookmakers to be less reticent. The way forward must be for more openness and critical evaluation, for us to stand up for what we do and believe in. Don't let bookart become rarefied and elitist, keep asking galleries to show your books, and send your work to be reviewed.

Since this is my opportunity to write what I want, I'd like to say something about my co-editors. John's imagination and resourcefulness as a practitioner, his experience of small press activities, has been essential to the belief that it would be possible to realise this project on a shoe-string budget. My correspondence with Stephanie has kept me entertained, with a variety of different letter-headings, insects are big this year. She revealed that there is something special and worthwhile about using pen and paper, echoed in her interview with Ian Breakwell, and I am convinced, that I'll never type a letter again. And then there's Stefan, who is truly inspired at getting people to work together. So thank you to the editors for all their hardwork, and to everyone who has in some shape or form contributed to the Yearbook.

Tanya Peixoto

BOUGHT AND SOLD IN '94

Mark Pawson

Mid-December and it's 'Books of the Year' time again, so here, in no particular order are 16 of my favourite publications by artists that I bought, sold or was given copies of in 1994. It's an eclectic list dating back over 30 years, with titles from the U.K, U.S.A, and Europe.

IMPRINTS 93 is a monthly series of artists' booklets, ranging from the totally unknown to Turner-Prize nominated artists. Edited by Matthew Higgs, there've been 15 so far, most are uniform A5 booklets, with greaseproof paper covers. *E1* by Hilary LLoyd is a sequence of typewritten diary accounts of the times she's been hassled by men in the street or on buses. *Some Broken Windows* by Matthew Higgs, is exactly that, a glazier's clip-art collection. Some issues deviate from the booklet format. Jeff Luke's piece is a replica bomb fusing device made of dowel and wire, and most recently, Jessica Voorsangers' *Partridge Family Christmas Album,* underneath a christmas tree christmas card. Free if you're on the mailing list. Back issues £2.00 each.

GRUESOME GREETINGS CARDS, bubblegum cards, Topps, U.S.A, 1992. I managed to track down these wonderful cute-sick cards, in the vein of the *Garbage Pail Kids and Toxic High School* sets which preceded them. Some are skunk-in-a-gasmask logoed scratch'n'stink cards, and they really do smell bad. In a significant policy change the artists and creators are all credited. Set of 44 cards $10, unopened box $6.

THE FRED, issues 9, 10, 11, London,1989. Poems, paintings, short stories, and adverts for nightclubs, fashion and hairdressers in 250 Gaultier suit pocket-sized pages, a total eighties time capsule, was it all only five years ago, it seems much longer...A bit of Gilbert and George and Duggie Fields in each issue. £2.00 each.

PORTRAIT OF ROBIN CROZIER, Sunderland, 1975. A postal art classic, for this project Robin asked a couple of hundred International Postal Art Superstars, none of whom he had met to send a portrait of him, then he redrew or made a portrait of all the portraits and put them all in this book. Robin gave out copies of this as prizes during the question and answer part of his Tate talk, for answers that seemed to amuse him the most. If you weren't there or didn't get a prize, expect to pay between £6 and £35, depending on where you shop.

WARRIOR FOR GRINGOSTROIKA, Guillermo Gomez-Pena, St Paul MN, 1993. I was so impressed with an extract from this, in an old issue of High Performance magazine, that I sought out a copy immediately. Gomez-Pena is a multilingual, multi-disciplinary, multicultural, multimedia social thinker and performance artist. This book collects poems, essays, manifestos, and performance texts from the last ten years, and I still haven't finished it. $14.

KATE MOSS, Jesus-la-Products, Welling, 1994. Appropriately atrocious print quality in this meditation on teenage 'supermodel' Kate Moss, as you turn the pages she disappears before your eyes, complicity of the gaze embodied in a limited edition, one copy for each year of age. 26p.

KINGDOM SCUM, Golden Asshole League, Eerie Materials, Richmond VA, CD in elaborate folder with booklet, stickers, etc. A visual and aural whacked-out-punk-sample-scramble-fest, this aesthetic even carries through to the cover stock, recycled paper with big chunks of almost legible newsprint in it. CD itself has an anti-technology rant printed onto it. $10.

REAL BULLSHIT, O.Milan 1993, Ben Vautier's Flux scandal sheet. "I thought that with such a title I could do some real bull-shit and say nasty things about nice people...made up of scribblings that date since 1990, I was just about to throw them into the waste paper basket when I looked through them and thought 'what's the use of being up to date if it's always the same shit' " Distributed at Ben's Tate talk, there are also some pirate copies around.

MANUAL OF BOOKBINDING, Thames and Hudson, London, 1978. I wanted to know how to do it, and couldn't be bothered to find a course, so I bought this and got sewing, making my first hardback book at home with zero expenditure on special equipment. £7.95.

LE DERNIER CRI, 6,7, Paris 1994. Endless graphic sex, death and violence by France's finest comic strip delinquents. Lines hacked with razors and the most luscious, curvaceous brushwork side by side. Tabloid, screened in 2 or 3 colours throughout, with foldout centrespread. .75ff/each

FLUXSHOE ADD END A, 1973, and Japanese Schmuck *8, 1976, both Beau Geste Press, Cullompton. Hey with these and the small stash of other *Schmucks* I've got in a box in my bedroom somewhere I could have my own l'il fluxbritannica show at home. £30, £27 and rising fast.

THE REALIST, 74, May 1967, New York. The most important ever issue of Paul Krassners paper, which proudly bears the motto: *Irreverence is our only sacred cow.* Includes 'The Parts that were left out of the Kennedy Book' where much of the JFK conspiracy theory comes from, including the 'story' of Lyndon Johnson 'interfering' with Kennedy's assassination throat wound. Wally Wood's timeless Disneyland Orgy centrespread is a classic in its own right. £2. each. After a 20 year hiatus *The XX Realist* is publishing again, current issues are $2 each.

ATLAS *4 ed Jake Tilson, London 1994, I seem to remember the first issue of Atlas announcing a quarterly publication schedule, now it seems to be every quarter of the decade. Theme for this issue is Beyond Drawing; found notes, drawings, alphabets notebook scribbles and sounds (on an 80 min CD), from a mathematician, musicians and artists around the world, trilingual, printed in Sweden and London £8.50.

MUSEUM FUR FOTOKOPIE YEARBOOK, Mulheim, Germany, 1993. Impressive documentation of the Trivial Machines show. Long essay, mainly in German, together with many illustrations of work, installations and machines. Extremely fragile embossed corrugated card cover creates an instant storage problem. No price.

NEW OBSERVATIONS *101, Copy Culture issue May/June 1994, New York. Important collection of articles examining the possibilities of widespread access to cheap photocopying, groundlevel information distribution, redundancy of copyright, zine and postering activity. Well researched article about Chester Carlson's (inventor of xerography) life and inspiration. I must declare a vested interest here, there's a pic of me on P.29. $5.

15 COWS, John Dilnot, Liverpool 1994, or to be exact, the mini-version of this accordion book that he sent to publicise the full-size screenprinted edition. This is a photocopied sheet with instructions on how to assemble the book, such a brilliant idea, maybe I'll use it. But will it translate into sales? Photocopy mini version, free, full size screened edition £5.

John Dilmot 1994

A REVIEW OF THE CHILDREN'S BOOKS DEPARTMENT AT

Dillons Bookstore

Mike Horl

My mother disapproves of anything revolutionary. "Why can't shops be the way they were?" she says. "Progress." I reply. "This is the end of the twentieth century."

Dillons is a very progressive, in fact some would say a revolutionary book shop. So revolutionary that it has a children's department for adults. In fact there is a Children's department, a Children's Spoken Word department, a Children's Video department and a Children's Educational Department. The logical arrangement and even the products themselves, have been selected to appeal not to the child, but to the person carrying the cheque book or credit card. Within each department are several sub-departments, offering a bewildering choice of gifts. The little chairs, toys and games at ground level are a welcome and relaxing addition to the shopping experience; their purpose is to keep the kids happy while the adults get on with the serious business of spending money.

Books are displayed from floor to ceiling, an effect which adds to the excitement of the whole visual experience. The most prominently displayed section is the one labelled *Picture Flats,* which would probably have been a better name for a 1960s housing estate. These books rely heavily on style of illustration to make instant appeal to the buyer. Familiarity and recognition also help to give reassurance as to the quality of the product, which explains the presence here of two books written by Graham Greene and illustrated by Edward Ardizzone. First published in 1974, *The Little Horse Bus* and *The Little Steam Roller* appear to come from an even earlier period. The former would definitely appeal to my mother. A friendly local grocer's business is threatened by the appearance

REVIEWS OF ARTIST'S BOOKS BY PUPILS FROM

KENTISH TOWN C OF E PRIMARY SCHOOL

Ella Finer, Pete Fraser, Chelsea Uddo, Katy Jockelson, and Katie Lovegrove.

The Upside Down Alphabet Book
Yaron Livay (Flying Sugar Press)

This is a brilliant funny little book - it is very clever. You'd think at first it was just an alphabet book but when you look closely you see the real cleverness.

Aeroplane Dreams
Phil Hopkins

This is strange but that is why I like it. It is very dream-like and calm. It is not a bedtime story type of book but much more a book that you'd look at for quite a while. There is more mystery than in a children's book.

Dreamwork
Mel Gooding & Bruce Maclean (Knife Edge Press)

At first glance you would think it wasn't much but once you had looked at it properly you would see the point.

I like the way it's laid out with the different ideas of people dreaming.

This is like someone's dreams and feelings. If you just look in it, it is like a baby has been born in it.

I had to look at it twice so I got to know the meaning.

It has such strange pictures in it. One picture is all red with blue lines on - what is that? It makes you really imagine.

of a big new grocer's shop across the street, "...it was owned by Sir William Popkins, who never came into the shop and never put lollipops in bags."

Contemporary stories tend to dwell on fantasy, using non-human characters. This avoids tricky problems of racial and sexual stereotyping, but it has developed a general formula based on untruths; animals, dinosaurs and monsters wearing clothes, riding motorcycles and speaking English. *Big Pig Gets Mad* , written by Willy Smax and illustrated by Karen Ludlow, is a book for the TV generation. Set in a Western town, the simple moral of "do not steal" is delivered in great style, with plenty of laughs on the way. *Hilltop Hospital,* by Nicholas Allan is set slightly closer to home and the situation is one with which we are more familiar. The characterisation is good and the animals very amusing, particularly Mrs Indianapolis, a pregnant elephant. Where the fantasy book is most successful, and most realistic, is in the hands of Jeanne Willis and Tony Ross. Their *Dr Xargle's Book of Earth Weather* uses the monster, or space creature as character, but we are not asked to believe it is human. The *Dr Xargle* series of books pokes gentle fun at the human race, particularly at everyday events we all take for granted. I'm looking forward to *Dr Xargle's Book of Earth shopping,* and so is my mother.

For very young children, Dillons Children's department features soft, squidgy, waterproof Bath Books, encouraging decadent habits of reading in the bath (no doubt while slurping on a long cold drink). Chunky, hard-wearing Board Books are an early introduction to the idea of book as object. Some pages are almost like building bricks! Again here, though, there seems to be a lack of new story-telling, with 'Thomas the Tank Engine' and friends being wheeled out for a shunt along a slightly different track.

Relying on name recognition and a formal, traditional appearance, rather than illustrative style, *Everyman's Children's Classics* are described by Dillons as "... handsome, full-cloth hard-cover editions - perfect Christmas gift books to treasure." These are books where reading comes first and the format demands that they are read with reverence.

Hovering somewhere between the picture books

Crocodile Puddles
Rob Hadrill
Even if you don't understand this book it's quite nice to read. But it is hard to work out what it's on about and the pictures are very strange.

Pyre
John Bently (Liver & Lights)

This is really secret and special being able to open a box and then having to move the lolly sticks to get to the book and then there still aren't words, it is all pictures. The 3D effect is the best bit...you want to get inside and touch it. What makes you want to touch it is the bit of fire sticking out...that's clever.

13 Potential Poems
Les Bicknell

My mum would love this. It's like a sculpture, so nicely presented. It is very crowded and there's lots to look at. It's a book for sharing and discussing. It is so imaginative what they've used...bits like old tickets, stamps, clay, wrapping paper, bits of crate. Having the soft see-through paper makes it feel precious. You treat these books well because you can see how much work they've taken.

General

These books are great — they are more visual than textual. Adults are mistaken when they think children must have words in books. You can read lots in these books even if there isn't a story because they tell you so much from the pictures and the added bits. They really appeal to me because they are illogical.

I think if you were used to fact books and things like that you might get impatient with these books because they're different and weird.

These books are really different — they really get you thinking. Practically everyone in my school loves these kinds of books and I think more things like this ought to be published. I think there are and the children's Education section is a free-standing collection of the best ever interactive devices for children, the *I Spy* series. Again we have here that instant parental recognition of a classic from the past, encouraging children to observe the world around them, to record information and experience the joy of collecting. Closer inspection reveals sadly that "Big Chief I-Spy," of the *News Chronicle Wigwam* has now been replaced by the disturbingly corporate "Michelin Man". *I-Spy Pets* and *I-Spy Classic Cars* I can understand, but *I-Spy Dinosaurs* ?

That word 'interactive' occurs again and again. 'Dorling Kindersley Action Packs' look interesting and each pack is said to contain more than 20 items, but without ripping them open and rummaging through the box it's difficult to confirm. The three titles in this series are *Pyramid*, *Night Sky* and, of course, *Dinosaur.* Dorling Kindersley also offer that other buzzword, 'Multimedia', in several CD-Rom packages. The obsession with animals reaches new heights, or depths, in *The Robot Zoo* subtitled "A mechanical guide to the way animals work", by John Kelly. I suppose it's slightly more civilised (and less messy) than an animal operating table.

Two collections by Walker Books stand out in the poetry section, *A Caribbean Dozen* and *The Walker Book of Poetry for Children*, although I can't help feeling they try a little too hard. I prefer the look and the content of *Ten Banana More !* "...poems for children by ten of the best poets writing today", illustrated by Brian Grimwood and published by Simon & Schuster Young Books.

Of the new hardback picture books, the most striking are *Ketchup on your Cornflakes*, by Nick Sharratt, a simple variation on the 'consequences' game, mixing and matching phrases, and *Miss Spider's Tea Party,* by David Kirk, whose glossy, surreal illustrations transport us into a gooey sickly-sweet world of cakes and jellies.

The most practical publications on view, for children, are the *Coping with ...* series by Peter Corey, illustrated brilliantly by Martin Brown. *Coping with Parents* is essential for dealing with readers of one book which definitely tells us this is the end of the twentieth century: *Parent Power – securing the best education for your child,* by Ray Massey, Education correspondent of the Daily Mail. This book is too many fact books published.

All these books are very strange in every way. I think they are very hard to understand. It seems better not to try to understand them but to think of them like a picture or a sculpture. If I owned one I might just have it on a shelf.

I think it's a nice idea to make books that are not only writing but sculpture and a piece of art to look at. I like the way that some of the books are very weird and you can look at them for ages because they are so detailed and eye catching.

They're improvised like we do in drama. They've made something out of nothing. But very carefully put together.

They're surprising - the recycling is brilliant.

I think adults would get impatient with these — but children would look more closely.

Not enough children's book are for looking.

When you get older you stop being interested in things that are just fun. When you go to a new place you start conforming.

Clever and witty books.

Makes you think, well, should books be just for reading? You read the telly in a way and that is all pictures. I mean you get clues about what will happen next. You do here really.

I think they're good because you can make up your own meanings. I bet we'd all say different things.

You have to treat them well. They make you realise how precious they are.

I'd love to try to make my own. If there was a workshop would it be just for adults?
I think 11-12 is the best age for starting on these sorts of books.

Editors Note: thanks, Kate Frood!

located in the Children's Education section but it's not the kind of book that twelve-year-olds will be rushing out to spend their pocket money on. The cover blurb reads, "Controversial classroom reforms have brought massive changes to the nation's schools. But have you kept up with the revolution?"

Mother?

The Little Horse Bus and *The Little SteamRoller,* written by Graham Greene and illustrated by Edward Ardizzone, published by Puffin (The Penguin Group).
Big Pig Gets Mad, written by Willy Smax and illustrated by Karen Ludlow, published by Orchard.
Hilltop Hospital, by Nicholas Allan, published by Red Fox.
Dr Xargle's Book of Earth Weather, written by Jeanne Willis and illustrated by Tony Ross, published by Red Fox.
Everyman's Children's Classics, various authors and illustrators, published by Everyman's Children's Classics.
I-Spy Pets, I-Spy Classic Cars and *I-Spy Dinosaurs,* published by Michelin.
Dorling Kindersley Action Packs, published by Dorling Kindersley.
The Robot Zoo subtitled *A mechanical guide to the way animals work,* by John Kelly, Dr Philip Whitfield and Obin, published by Hamlyn.
A Caribbean Dozen and *The Walker Book of Poetry for Children,* published by Walker Books
Ten Banana More ! ...poems for children by ten of the best poets writing today.", illustrated by Brian Grimwood, published by Simon & Schuster Young Books.
Ketchup on your Cornflakes, by Nick Sharratt, published by André Deutsch.
Miss Spider's Tea Party, by David Kirk, published by Virgin.
Coping with Parents by Peter Corey, illustrated by Martin Brown, published by Hippo (Scholastic).
Parent Power – securing the best education for your child, by Ray Massey, Education correspondent of the Daily Mail, published by Harmsworth Publications.

Mike Horl

Dear John, Here is a little bit that you may ignore if you like:

Most art today, because it is a currency, is behind glass or must not be touched, but the art book is a form of expression – working at the magical distance between eyes and fingertips – which ~~brings back the tactile~~ involves the sense of touch. It is also as intimate as a letter and may involve the element of surprise. ~~It~~ It is theatre, but you may touch the actors!

Art books could be the biggest thing in this country since pottery!

Hope you did well. A couple of old prospectus' etc enclosed

Good to see you,

Best wishes

Andrew

Many bits I love in yours –

Nylon raggled-bags with pantaloons

wedhe on eyes

It darkly flapped

the red thorn and green leafery

by Andrew Lanyon,

BROUGHT TO BOOK

Stephanie Brown

Talks to Ian Breakwell and Paul Hammond.

Brought to Book: The balance of books and life, was compiled by Ian Breakwell and Paul Hammond to counter the myth that books are a closed, escapist world. A wide range of contributors reveal how personal experiences of a comic, tragic, erotic, mysterious and bizarre nature have forged extraordinary connections between books and the world around them.

In working on the book, Breakwell and Hammond also worked closely with the unexpected, and coincidence and serendipity played a major role in its production. Not the way a mainstream book is usually put together, and yet *Brought to Book* was published by Penguin, probably the best known of conventional publishers.

Coming from a visual arts background, and using creative processes such as chance and collage, Breakwell and Hammond demonstrate how artists can move from the world of artists' books and small presses onto the shelves of W.H. Smith's, Waterstones and Dillons without compromising their ideas or losing control over their work.

Stephanie Brown talked to Ian Breakwell and Paul Hammond at Penguin's Kensington headquarters on 29 September 1994, the day their book was launched at the I.C.A.

SB: Sorry, I forgot to turn the tape on. Look 'No Smoking' and no ashtrays either. How did you know who was likely to have a book story and how did you go about finding contributors?

IB: The methods we used were not those that conventional publishing would adopt to put an anthology together. We operate from a small press background and for some time Penguin had difficulty in understanding our methods. But they knew from *Seeing in the Dark* (A Compendium of Cinema-going also co-edited by Breakwell and Hammond, Serpent's Tail 1990) that our methods worked and that's what they were interested in. We could get a type of anthology they couldn't get. It was done by personal contact and building up a network, of pen-pals almost, starting with my address book and writing off to people who then contacted others. All letters were hand-written and individual and all replies were personally acknowledged. We didn't use computers or those falsely individualised word-processor letters. It was all very *ad hoc,* following our noses, like a paper-chase, one person leading to another. There could be a big ideological argument about word-smithing, of using pen and paper. It's closer to being an artisan.

SB: It's more organic as well, just as the content of the book evolved in an organic way, by direct contact with people rather than pre-packaged ideas assembled in a sterile way.

IB: Because it's organic, which is exactly the word I was searching for and the right one, it leads into areas which were unexpected to us, and the theme and direction of the book started shifting and taking us by surprise. We discovered things of which we had no real conception. We also thought we might come across people who actually loathed books, but we didn't.

PH: There's no such word as book hater, did you know that?

SB: No. Come to think of it bibliophiles are common enough but I've never heard of a bibliophobe. I suppose they're selective. Savonarola, the Inquisition and the Nazis were all selective book-burners rather than burning books *per se*.

PH: Yes. We did though, come across a few incidental stories to do with things like allergies, which put people off books.

SB: I see you're flicking your ash into a piece of paper, I'm using my shoe.

IB: Going back to contributors, we had a pre-publication drinks do at Compendium on Monday. Six contributors came up and introduced themselves and I guessed four of them before they said their names. It's like you know the person. Our knowledge of them is quite deep and personal even without knowing them physically. We know their written voice. We ask people for something: they think about it and then present us with something, so it's very personalised but without any physical meeting.

PH: We do strike this personal note with people and we have to have their trust because we don't send them copy. They don't see their piece until they get the book in their hand.

SB: So you make it clear to them that you might edit or modify their copy?

IB: Yes. There is, for example, a major problem regarding unwitting repetition. Someone sends us a story unaware that someone else has written us something very similar. Sometimes we've had on two successive days, virtually the same story from two people on opposite sides of the world.

SB: Amazing! Amorphous resonance.

PH: Yes, in one week we've had six separate contributions, three of which end with the same sentence. Even if you spread these through the book , it looks like the classic word-processor sloppiness. So you have to change some. Also, some are just too wordy, but might contain one terrific thing - a sentence or so - so we take that. A lot of these things found their way into the marginalia that we used in the design of the book.

SB: I was going to ask you how the marginalia came about as it's effective both as a design device and as a foil for the more substantial texts.

IB: The marginalia is 70 words maximum and dotting them about, like asides in a play, works very well. Apart from out-takes they give the chance for someone with a little, punchy anecdote to be in. They also allow for forms usually kept out of mainstream books because of their shortness, but we can handle that, and also the visuals, within our design format.

SB: How did you select the visuals? Clearly you came across these in a different way to the texts.

IB: It was a separate part of the research, unless someone sent us a drawing or a photograph. The sculptor, David Mach for instance, said he didn't have a book story but he sent us a photo of a Rolls Royce he'd made out of remaindered books. We use the photos as we use the marginalia - alongside the text in a way that sparks off against it. When we had around two-thirds of the text, we trawled the picture libraries, looking, not necessarily under 'books', but under the subjects of the stories we were getting. We got all the stuff around us and assembled everything using a collage-type approach. A broad framework emerged whereby we started with birth, and ended, not necessarily with death, but something leading towards it. Things grew together - pieces about emigration, a piece about absence...

PH: As Ian says, it starts to grow itself, after that it's just the graft of sorting out bunched up photos so there's a better spread through the book.

IB: This goes back to our previous discussion about computers. In our ignorance, we work using intuitive methods rather than placing information in a certain sequence. The other thing to consider is that we both trained as, and still are, visual artists. Paul still paints, and has done lots of collage, so have I. I've also done films and edited videos so I also know about assemblage and montage. So the book's put together, not necessarily in a literary way, but much more in an artistic way.

SB: So that's why it's appropriate to discuss it in

the Yearbook. The whole thinking behind it and the process of putting it together, can be closely associated with the creative processes behind many artist's books. Have you encountered problems using that type of approach but working with a set-up like Penguin, which presumably has preconceived ideas about how to proceed?

PH: Can I just say one thing before looking at that. Penguin approached us on the strength of the cinema book. Our track record there meant that they accepted the book on the strength of the concept alone.

IB: I remember when my *Diaries* which were in lots of small press editions, came out in paperback for the first time. One of the small press editions was reviewed, by Paul in fact, in *Art Monthly*, so when it came out in paperback Peter Townsend said "We'll definitely review that."It went into the Artist's Book section where they took this very snotty view...

SB: Ah, that would be Cathy Courtney.

IB: Yes it was, and she was very snotty, as if just making the move into mainstream publishing was somehow distasteful, and some kind of cop out. The content of course, was identical - completely unaltered. Anyway I always wanted them in trade paperback because they could go far beyond the art world and reach anybody. The same week I also published an original *Diary* folio, in fifty copies - it's horses for courses.

SB: Horses ok, but what about problems with the Penguins?

IB: Although you might get difficulties in coming from an art background, it's a question of saying, "What can we do with commercial publishers that we couldn't otherwise do, what are the advantages, and how can we work with that?" We can reach the public for a start, but Penguin has no other book in its stock like this, so it's going to be an odd one for them - rub against the grain of conventional publishing. A bit of a problem.

PH: That's heightened by the fact that they didn't know where to put it in their trade catalogue. In fact they had it under reference and now it's in literature / literary criticism.

IB: So there are problem areas, but the plus is that the book goes to places where the distribution set-ups of small publishers could never, ever reach. That is the one big problem with artists' books and small presses, always has been, always will be...

SB: Distribution.

IB: Poor distribution. Whereas this will go into W.H.Smith's...

PH: It won't only be on shelves in mainstream bookshops, it will also be displayed, they usually display new Penguins.

IB: Yes, and therefore it's bound to reach people who would never normally read exactly the same thing if it was in a small press edition. Tescos shifted an enormous number of paperback copies of a supposedly obscure Marques book, but there's absolutely no way you could get a small press book on supermarket racks. That's the plus side, and there was no editorial interference or censorship. Another plus is that contributors to our book get no royalties. With some contributing a few lines , and others five thousand words, it would be impossible to calculate royalties, so they're getting nothing and royalties are being donated to the RNIB for Braille and talking books. But Penguin means something to people; contributors appreciate being associated with Penguin, the pioneers of putting books within the reach of everyone on the street.

SB: This is a democratic ideal of course, and it's significantly antithetical to the aims and values of many makers of artist's books.

IB: I sometimes feel the artists' book world is very inward looking. The impression of having to steer clear of the tainted world of commercial publishing is almost endemic and it can be terribly precious. Yet, the history of artists' involvement with books through the ages doesn't bear that out. Especially those artists who are also writers. *They* didn't see it as a rarefied thing.

SB: Blake for instance, putting together books on the kitchen table, trying to get them out cheaply to people, although it turned out nobody wanted to buy them. The point you make about preciosity and the rarefied thing is true, particularly in the overlap of artists' books with the more twee and tasteful end of Fine Presses.

IB: I think there's a fault amongst the artists of not conceiving the work as being possible beyond a little coterie. But there's also a problem amongst publishers which I hope is breaking down a bit. Certainly Serpent's Tail for instance, consistently publish people who do other things than write, performance artists, film-makers or whatever. The more 'literary' publishers haven't really registered that cross-over and are often unaware of the fact that a lot of artists' work could be commercially published, working with a good editorial team and a good team of designers. A good example is Glen Baxter. Effectively what he's doing now with Thames and Hudson and Bloomsbury, is what he was doing twenty-five years ago - self-published, stapled together things. He hasn't changed it basically, but after a massive amount of small press stuff, somebody notices, and having that experience behind him means he's going to be intransigent when it comes to how big publishers present his work. So small presses are ahead of the game to this day and it's a question of how aware publishers are of what's going on. You have to have half-way publishers, like Serpent's Tail in England and City Lights in the States - they bring out books by Lucas Samaras, Karen Finlay etc. in paperback.

SB: Perhaps mainstream publishers will become more aware of the possibilities as CD-Rom and so on progressively replace books as the servile carriers of information.

IB: Yes, but I don't think anybody's yet thought of these technological tools as a means of changing *writing*. Take William Burroughs. For years and years he was trying to work like a visual artist with collage - banging it out on an old Olympia typewriter. He's too old now but actually what he needed was the world's first word-processor. You can cut up, collage, shift things, and he would have found a way - without scissors and paste - to produce a different kind of fiction. Instead people just use these things to speed up and bulk out conventional writing.

PH: I think there are visual artists across all media, and using all kinds of sophisticated methods - not all, but some of them - who would definitely cross the bridge into mainstream publishing if they realised it themselves.

IB: Especially with the visual/verbal book becoming much more common. Sometimes though, when we asked artists to contribute to the book, they sent a load of slides in. We can't use them, and the artists hadn't thought about them in terms of the printed page, they were thinking of it as some kind of catalogue. It's not an art book with a capital 'A'. Also they needn't have sent a picture. Why not write something? Some say, "That's not what I do" and that's it. Others have thought about it, done it, and discovered they can write as well. I always used to loathe the Clement Greenberg thing - the promotion of the artist as a sort of verbal illiterate. Pollock for example.

PH: He wasn't interested in artists speaking for themselves because he wanted to speak for them. I think though, that there's work by plenty of artists that could cross into trade paperback, and plenty that definitely couldn't. For example anything wholly dependent on colour, which would be economically difficult, although that may change too.

STORIES OF THE EYE

BY IAN BREAKWELL

In a three-part series, Ian Breakwell rediscovers the fantastic and sometimes scurrilous literary output of some celebrated names. Parts 1 and 2 were first published in *Artscribe.* However, this is the first time all three sections have appeared together. In the first instalment, Breakwell looks at the art in language of Duchamp, Kubin, Klee, Beardsley and Dali.

Story One

Although poles apart in all other respects, William Blake and Marcel Duchamp both concluded that, as need arose, the artist must become a writer. A multifarious visual artist and cunning linguist, Duchamp extended *The Large Glass* into a labyrinthine shaggy dog story of interdependent visual and verbal imagery. His playful awareness of language was embodied in a lifelong output of verbal constructions, puns, spoonerisms and "texticles" which ran parallel to his installations and ready-mades.[1]

The dispassionate, games-playing Duchamp acknowledged a kindred spirit in the painter, graphic artist and writer Paul Klee, noting that: "He has so much to say that a Klee is never like another Klee." Such oscillating fecundity is potentially unhinging, but Klee was an orderly man, who knew that if you spend every day conjuring up fantastic imagery which transcends mundane reality, then for sanity's sake it must be counterbalanced by ritualised routines. "I once led a restless mode of life, until I acquired a natural base that enabled me to abandon that mode...My vital functions were locked up somewhere in a strong box, well preserved. An ear applied to it could hear the pulse ticking softly..."

Klee's poetically logical analysis of creative processes in *The Pedagogical Sketchbook* and *On Modern Art* takes a more discursive form in his *Diaries,* which incorporate travelogues, musical appreciation, playlets, anecdotes and jokes.[2] Klee is revealed as a comedian, haunted by an awareness of life's darkness, which he alleviated through elaborate compositional games. Klee was an illuminating writer, but only one of many visual artists skilled with words.

Klee exchanged pictures with Alfred Kubin, who in 1907, after his father's death, found himself unable to paint or draw. In twelve possessed weeks Kubin wrote the fabulous novel *The Other Side.* In it an artist (obviously Kubin) receives a mysterious summons from a forgotten school friend, Claus Patera, to live in the Dream Kingdom, set in the depths of Asia, of which the frustratingly inaccessible Patera is creator and ruler. Dreamland turns out to be a flyblown dump where time has been put back 50 years. An American canned-beef tycoon, Hercules Bell, attempts to overthrow Patera and introduce big business. The ensuing struggle sees the gradual disintegration of Dreamland's civilisation ending in cosmic cataclysm. *The Other Side* anticipates Kafka in its hallucinatory mix of sordid realism and comically unnerving nightmare:

"An elderly fellow with an abnormally big torso and short legs approached me; he was naked except for a pair of greasy duck workman's trousers. He had two long vertical rows of nipples - I counted eighteen. With a snort, he inflated his lungs, filling first the right and then the left side of his chest, and then with his fingers on the nipples played the most beautiful harmonica pieces. As he expelled his breath he moved to the rhythm of the melody like a dancing bear."[3]

In his autobiography, Kubin describes how *The Other Side* was written "in an extraordinary state of mind that was literally comparable to intoxication. I felt as though actually possessed of true clairvoyance."[4] Visionary in both form and content, the escalating sequence of disasters which push Dreamland towards apocalypse are startlingly prescient of later works as various as Alfred Hitchcock's film *The Birds* and William Burroughs' sci-fi celebrations of anarchic chaos.

"The melancholy cawing of crows drew my attention. The big black birds were sitting close packed in long rows on the brick works. Occasionally the whole flock would rise and perform the most precise manoeuvres in the air. In the direction of the river, the sky was still reddened by the burning mill."

"A small hospital revolution broke out. The enraged patients quickly reached an understanding among themselves and fell on Doctor Lampenbogen...A miserable woman with a broken jaw painstakingly poured chloroform drop by drop over the doctor as he lay groaning in his fat...They skewered Lampenbogen on a gas-pipe - a long and tedious operation for the weakened patients. The orderly made a fire so as to obliterate the traces of the outrage. So Lampenbogen ended his existence as a roast *en brochette,* and what's more, badly done; his back was almost raw, barely warmed through; the belly on the other hand, was completely charred. Only his sides were nicely crisped."

"The swamp was devouring the station. The building itself had tipped over, the platform was covered with slime and bullrushes; through the rotting doors the morass crept into the waiting room, the melancholy song of toads resounded from the benches and cushions. Over the buffets crawled salamanders and grubs."[5]

Ecological ruin, the collapse of social services, inner-city riots and unstoppable global carnage are all prefigured in Kubin's astonishing book. *The Other Side* is the key to Kubin's art, for he illustrated its teeming verbal imagery in hundreds of later drawings, one being of an artist both he and Klee admired: Aubrey Beardsley, pictured on his deathbed, coughing out his last tubercular gasps. One wouldn't have thought poor Beardsley had

much to laugh about, yet his novel *Venus and Tanhauser* is a hilariously lewd spoof of the amorous encounter between the Goddess of Love and the legendary knight. Venus' elaborate dressing and toilet preparations culminate in masturbating her pet unicorn and lapping up his sperm from the floor. After an energetic bout of fucking nearly overturns a carriage, the weary Tanhauser walks by the lake to get his breath back.

"Sometimes the lake took fantastic shapes, or grew to twenty times its size, or shrunk into a miniature of itself, without ever once losing its unruffled charm, its deathly reserve. When the water increased Tanhauser was very frightened, for he thought how huge the frogs must have become. He thought of their big eyes and monstrous wet feet, but when the water lessened, he laughed to himself, whilst thinking how tiny frogs must have grown. He thought of their legs that must look thinner than spiders', and of their dwindled croaking that never could be heard."[6]

Admittedly I favour Beardsley's perversely camp version of the myth: my idea of insufferable purgatory would be compulsory attendance at Wagner operas. There is no denying, though, Wagner's influence on artists' writings: "Tristan and Isolde" was evocatively used in Luis Buñuel's film *Un Chien Andalou,* co-scripted by Salvador Dali, who up-dated the same legend of love-in-death in his vast baroque novel *Hidden Faces,* a convoluted tale of a group of eccentric French aristocrats at the time of the Paris riots of 1934 to the last days of the Second World War. [7]

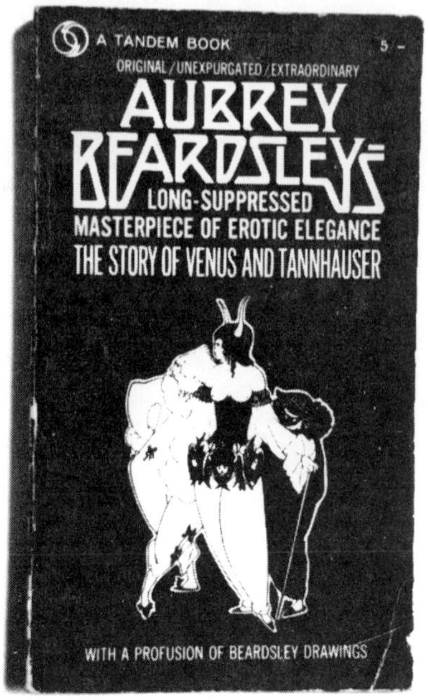

Dali's vision of Hitler as the supreme masochist, embarking on a war in order to lose heroically, provides the novel's Wagnerian finale. His ambivalent obsession with the Führer reveals that his political opinions were synonymous with his sexual fantasies: "I was fascinated by Hitler's soft, round back always so tightly encased in his uniform. The softness of the Hitlerian flesh squeezed into the military tunic brought me to a state of ecstasy."

Dali's literary style, a verbal equivalent of the photographic hyperrealism of his paintings, juxtaposes interminable pages of turgid romanticism with biting satire on the American *nouveaux riches* - the very patrons he was later fawningly to court - encapsulated in the image of a jaded flying ace caressing an heiress's thighs with

his shadow. *Hidden Faces* is Dali's delirious swan-song before his long decline into vacuous self-promotion.

Dali said: "I hate simplicity in all its forms" often confusing obsessive detail with the precision he sought "to materialise the images of concrete irrationality." Thus, his supposed autobiography , *The Secret Life of Salvador Dali,* pre-dating *Hidden Faces* by two years, is a tour de force of manic narcissism, rhapsodising his excremental fixations in a torrent of verbal diarrhoea. Dali's self-justifying confessions of his unashamed careerism and cultivated abnormalities are at best an outlandish self-analysis far surpassing conventional psychoanalytical writing, at worst a tiresome gossip column wherein art becomes mere kitsch fashion and publicity.

"(At) the first 'surrealist ball' society

women appeared with their heads in bird-cages and their bodies practically naked. Others had painted on their bodies frightful wounds, cynically slashing their beauties and transpiercing their flesh with a profusion of safety pins. An extremely slender, pale spiritual woman had a 'living' mouth in the middle of her stomach gaping through the satin of her dress. A man in a bloody nightshirt carried a bedside table on his head, from which a flock of multicoloured humming birds flew out at a given moment..."[8]

Showman and shaman, Dali was the forefather of Andy Warhol, Joseph Beuys and Jeff Koons, and funnier than all three, though the jokes eventually wore thin.

The years from 1952 to 1963, when Dali achieved world-wide notoriety while simultaneously forfeiting the respect of fellow artists, are chronicled in the modestly titled *Diary of a Genius,* notable for its definitive treatise on "The Art of Farting".[9] The timbre, resonance, trajectory and aroma of every conceivable kind of fart, from the "Double-Barrelled" to the "Silent Creeper" is meticulously analysed and set in context according to circumstance, climatic conditions, geography and provenance, from the marjoram-scented emanations of Shepherdesses to the terrible anal assaults of Fencing Masters. The etiquette of farting, fart music, and the medical potential of "Spirits Of Fart As A Cure For Freckles" are mulled over in bowel-tickling detail, concluding that Farts Are Of Benefit To Society".

Dali's subsequent sell-out to Café Society and Roman Catholicism enraged the surrealists who anagrammatically disowned their most famous painter as AVIDA DOLLARS. But, as we shall see in the next instalment, those Surrealist painters and sculptors who Dali deserted also happily ignored his own *Diary* maxim: "Painter you are an orator. So paint and keep quiet.

(Notes: Where available the date of first publication or translation is given in brackets. Prices are notified on books still in print.)

1. Marcel Duchamp, *The Essential Writings of Marcel Duchamp,* Thames and Hudson, London, 1975.

2. Paul Klee, *The Pedagogical Sketchbook,* Faber & Faber, London, 1966 (1948) £5.99; *On Modern Art,* Faber & Faber, London, 1966 (1948) £5.99 *The Diaries of Paul Klee 1898 - 1918,* University of California Press, California, 1968 (1964) quoted pp156, 316.

3. Alfred Kubin, *The Other Side,* Penguin Classics, London,1973 (1909) quoted p.132.

4. Alfred Kubin's "Autobiography", compiled over the course of his life is included in the Penguin Classics edition of *"The Other Side," op. cit.*

5. Kubin, *op. cit.* pp 196, 204-6, 206.

6. Aubrey Beardsley, *"The Story of Venus and Tanhauser",* Tandem, London, 1967 (1907) p.144.

7. Salvador Dali, *Hidden Faces,* Peter Owen, 1990 (1944) £8.95.

8. Salvador Dali, *The Secret Life of Salvador Dali* Vision London, 1981 (1942) p.337.

9. Salvador Dali *Diary of a Genius,* Hutchinson, London, 1990 (1964).

Story Two

Leonora Carrington and Meret Oppenheim: Artists who turn pen and ink to the deviant pursuit of literature.

> "The visual arts had no place in the original blueprint for the surrealist movement, which was dominated by writers avowedly dedicated to the revolutionising of society at every level."[1]

But every level includes visual expression, and when the painters and sculptors swarmed on to the literary vessel like pirates, Captain Breton, discreetly valorous, welcomed them aboard. This conjoining of seeming opposites expanded the group's verbal potential far beyond anything that the writers, left to themselves, might have envisaged. Every conceivable form of literary

expression was subsequently explored: from the intermingling of writing and drawing formerly typical of schizophrenic art, through novels, plays, manifestos, critical essays, aphorisms, anagrams, puns, film and radio scripts, cookbooks and unclassifiable prose texts of all kinds.

The fluid equation between visual and verbal means is particularly pronounced among women Surrealist artists. Meret Oppenheim's style-hopping output in a diverse range of media: paintings, drawings, sculpture, theatre events, masks and costumes, strange jewellery and furnishings, as well as famous objects such as the *Fur Cup, Saucer and Spoon,* exemplify her belief that "Every idea comes with its form." Thus some ideas become verbal avowals: "I love those cold flowers of laughter best and their nodding shadows that shine in the dark."[2]

Oppenheim the dilettante, touching everything lightly, is touched discriminately in turn, unlike the magpie that sees only glister:

> "The voyage has lasted more than a thousand years. People swim around their ship like sharks and the sea is red with blood. Only the brown dogs stick their heads over the sides of the ship. They hold little knives in their jaws and drop them from time to time on the people below. The knives impale the sun which lies deep at the bottom of the sea. The sun grows little fins."[3]

In 1939 Oppenheim wrote little stereoscopic sentences to be viewed through an optical device enabling each eye to read its own sentence: three dimensional writing, a witty verbal objectification of the collage principle central to Surrealist art and writing. In the case of Leonora Carrington, whose paintings and writings are packed with the most unlikely juxtapositions - of character, circumstance and plot - her singular ability is to weld these collage elements into sequential narratives of dream-like logic.

Carrington was born in 1917, in Lancashire daughter of an English textile tycoon father and an Irish mother. Parental efforts to groom headstrong Leonora into a high-society housewife after debutante presentation at court, proved fruitless. In 1937, aged 19, she met 46-year-old Max Ernst in London, eloped with him to Paris and became a leading figure in the Surrealist movement. Escaping from the Nazis she settled in Mexico, where she delivered an outpouring of paintings, drawings, stories and plays which has continued to the present day.

Carrington's lifelong pursuit of magic, occult and arcane studies, immersion in the teachings of Gurdjieff, Tibetan Tantric and Zen Buddhism combined with the Celtic legends and faery literature learned from her mother, influenced all her work. In the wrong hands this source material would produce yet more fey Tolkienesque fantasy. But Carrington is an alchemical artist, and from potential dross she has distilled pure gold, especially in her writings. Carrington's love of cookery and magic potions often intermingles:

"Dominique prowled around the table with soft steps, serving delicate dishes, a plump fat chicken with stuffing made with brains and the livers of thrushes, truffles, crushed sweet almonds, rose conserve with a few drops of some divine liqueur. This chicken, which had been marinated - plucked but alive - for three days, had in the end been suffocated in vapours of boiling patchouli: its flesh was as creamy and tender as a fresh mushroom." [4]

Self reference abounds in her stories, notably in the recurring figure of the wild woman with a feral kinship for animals:

"Her name was Virginia Fur, she had a mane of hair a yard long and enormous hands with dirty nails...Of course, she had to put up with being insulted by the cats sometimes, but she insulted them back just as loudly and in the same language...Every night she went out on her wheel to hunt; whatever their respect for her the mountain beasts didn't let themselves be killed as easily as that, so several days per week she was forced to live on lost sheepdog, and occasionally mutton or child." [5]

Though steeped in mysticism she maintains a level-headed contempt for gurus:

"'D is here this evening,' the young man continued. 'He is a magus and I am his pupil. Look, there he is, near that big blond dressed in purple satin. Do you see him?' I saw a man of such neutral appearance that he struck me like a salmon with the head of a sphinx in the middle of a railway station. The extraordinary neutrality of this individual gave me such a disagreeable impression that I staggered to a chair." [6]

The narrator's mock melodramatic reaction to the grey eminence rings a note of surprise rare in Carrington's stories, usually characterised by a matter-of-fact tone which authenticates the bizarre scenarios. This deadpan documentation is taken to its extreme in the autobiographical *Down Below,* which chronicles her distraught madness triggered by Max Ernst's wartime internment, and her committal to an asylum in Madrid where she was strapped down, force-fed and injected with the shock-inducing drug Cardiazol. The nightmare is recounted retrospectively in a rational, objective tone as if describing another, separate being. The effect is terrifying. Her personal experience of the anguish of going stark staring mad contradicts the Surrealists' identification of insanity as the ultimate irrational journey to a state of creative blessedness.

Down Below reveals madness to be a wretchedly unwilling loss of individual self which culminates in the sufferer degenerating into a non-person, a cliché, convinced like innumerable other misguided patients that she is Christ on earth in the person of the Holy Ghost. The Monarch of Misrule becomes dismal Jesus. Cul de sac. Carrington's portrayal of the numbing banality of insanity is deeply disturbing:

"Three days perhaps after my second Cardiazol injection, I was given back the objects which had been confiscated on my entering the sanatorium and a few others besides. I realised that with the aid of these objects I had to set to work, combining solar systems to regulate the conduct of the world. I had a few French coins, which represented the downfall of men through their passion for money...My red and black refill pencil (leadless) was Intelligence. I had two bottles of eau de cologne: the flat one was the Jews, the other cylindrical one the non-Jews. A box of Tabu powder with a lid, half grey and half black, meant eclipse, complex, vanity, taboo, love. Two jars of face cream: the one with the black lid was night, the left side, the moon, woman, destruction; the other , with a green lid, was man, the brother, green eyes, the Sun, construction. My nail buff, shaped like a boat, evoked for me a journey into the unknown, and also the talisman protecting that journey: the song *El barce velere.* My little mirror was to win over the whole."[7]

In 1974 Carrington pulled all the disparate strands of her work together into a masterpiece: the novel *The Hearing Trumpet*. The story is narrated by 92-year-old Marian Leatherby, unwilling inmate in an old people's home regulated by the Inner Christianity beliefs of the insufferable Doctor Gambit. After the poisoning of one of the old ladies, whom post-mortem reveals to be a man, the women rise up in a cabbalistic coven against Gambit's regime and unwittingly set off a cosmic cataclysm which plunges America into a second Ice Age. By suitable paradoxical proxy, this realises Marian's lifelong dream of living in the snow-capped land of sleighs pulled by wooly dogs. As the book's last line and motto says: "If the old woman can't go to Lapland, then Lapland must come to the Old Woman."

A *tour de force* of comic writing, combining startling symbolic fantasy with a gripping narrative, *The Hearing Trumpet* is an outrageously optimistic celebration of extreme old age. Finally *Down Below's* traumatised exploration of the involuntary loss of self is drolly exorcised:

"I tried to nod and move away at the same time, but my knees were trembling so much that instead of going towards the staircase I shuffled nearer and nearer the pot. When I was well within range she suddenly jabbed the pointed knife into my backside and with a scream of pain I leapt right into the boiling soup and stiffened in a moment of intense agony with my companions in distress, one carrot and two onions.

"A mighty rumbling followed by crashes and there I was standing outside the pot stirring the soup in which I could see my own meat feet up, boiling away merrily as any joint of beef. I added a pinch of salt and some peppercorns then ladled out a measure into my granite dish. The soup was not as good as bouillabaisse but it was a good ordinary stew, very adequate for the cold weather. From a speculative point of view I wondered which of us I was." [8]

NOTES

1. Ed. Lucy Lippard, *Surrealists on Art* , Prentice Hall, New York, 1970.

2. Meret Oppenheim, c1933 cited in Bice Curiger, *Meret Openheim*, Parkett, Zurich - Frankfurt, New York, 1989. p.110.

3. Meret Oppenheim c1935 ibid. p.110.

4. Leonora Carrington, "Monsieur Cyril de Guinder", c1937, Originally published in *Pigeon, Vole,* Editions le Temps Qu'il Fait, Cognac, France, 1986. Reprinted in *The Seventh Horse* (1939) Virago Press, London, 1989. £5.50.

5. Leonora Carrington, "As They Rode Along The Edge" c1937, Originally published in *Pigeon, Vole,* Reproduced in *The Seventh Horse* (pp 3-4).

6. Leonora Carrington, *"The Neutral Man"*, c1950 originally published in *Le Surrealisme Même,* no.2. Paris, 1957. Reprinted in *The House of Fear,* (p147). £10.99.

7. Leonora Carrington, "Down Below", originally published in *ITT* no. 4, 1944. Reprinted in *The House of Fear,* (pp195-6).

8. Leonora Carrington *The Hearing Trumpet,* Routledge and Kegan Paul, London, 1977. Originally published by Editions Flammarion , 1974.

Story Three

Francis Bacon's assertion that Peter Owen is "the only publisher in England who produces anything interesting" would, justifiably, have been challenged in the past by the likes of Jonathan Cape and John Calder, and in the present by Serpent's Tail. Yet visual artists retain special affection for Owen, who in forty years of adventurous independent publishing, has consistently commissioned high-quality illustrations and also published the writings of artists such as Marc Chagall[1], Roland Toper[2], Jean Cocteau [3], Yoko Ono [4], and Giorgio de Chirico [5].

Hebdomeros, de Chirico's only novel, is a plotless prose poem verbally paralleling his paintings' enigmatic world of pellucidly lit arcades and squares:

> "Now the hours had passed by, slowly, fatally, as all hours pass. The sun was high in a cloudless sky, but veiled in a soft mist which heralded the approaching summer. Not a breath of wind...The shadows on the sundial marked noon."

But unlike the paintings in which, like cities depopulated by the Neutron Bomb, direct human contact is lost and people are replaced by statues and shadows, the novel teems with humanity, mordantly observed by the daydreaming outsider Hebdomeros:

> "Like a Colossus of Rhodes infinitely enlarged in dream, his legs apart and his feet touching different regions; between his left toes Mexican Bandits pursued each other like famished wild beasts around rocks overheated by the canicule, while his right foot, up there, trod the immaculate regions, among white bears, obese old men with weasel-like profiles shaking their heads in front of glaciers that were slashed, crenellated and jagged, like ruins of famous cathedrals, destroyed by cannon fire, and on the thresholds of stinking huts, woven from sealskin, Eskimos trussed up in furs politely offered their wives to excite explorers."

Permeated with atmospheric melancholy, yet lightly comic, *Hebdomeros* is the hinge between de Chirico's period of lyrical metaphysical paintings and the heavy-handed classicism he subsequently went to preposterous lengths to justify in *The memoirs of Giorgio de Chirico*[6] a paranoiac tirade of unreserved egocentricity rivalling Dali, for whom, ironically, he reserves special scorn.

Like a statue from his paintings, de Chirico's awesome presence overshadowed his brother Andrea, who changed his name to Alberto Savinio. A distinguished painter and composer, Savinio's idiosyncratic writings have been championed by the Marlboro Press, which specialises in contemporary Italian translations. *Capri* [7] is an amiable sketch of the Mediterranean Island famously associated with Tiberius and Gracie Fields, punctuated with imagined meetings with mythological figures and vertiginous inclinations to jump off the isle's scenic precipices. These suicidal urges during a jolly day trip typify the panic undercurrents which lie between the lines of Savinio's rational and measured prose.

Speaking to Clio [8] the journal of a trip to the Etruscan burial places, is dedicated to the muse of written history, which "collects our actions and gradually deposits them in the past, and allows us to rediscover ourselves each morning in a fresh condition, free from the past," and thus "give form and shimmer to the continuous present in life." The book celebrates the art of strolling, and asks: "Why do men in cities walk in such a hurry? To escape the thought of death." So, Savinio and his fellow death tourists wander from tomb to tomb, mockingly confronting the grim reaper, while "Over us fly huge, heavy, featherless birds made only of grey and leaden flesh, and on this land consecrated to death, their shadows pass like the shadows of airplanes."

Savinio's meandering erudition reaches sublime heights in *Operatic Lives* [9] , thirteen 'Libretto' biographies of illustrious persons as varied as Verdi, Nostradamus, and Carlo Collodi, the creator of Pinocchio. A chilling account of Isadora Duncan's death by car after her son suffered a similar fate is intercut with typical Savinio digressions:

"One of the few illustrious Frenchmen of the time who did not seek labial contact with Isadora's skin was Eugène Carrière. A painter of opacity and fog, Carrière was also a pure and beautiful soul. He lived a humble and saintly life in a top-floor studio among his books , his family and his friends. 'When coming into his presence' says Isadora, 'I felt as I imagine I would had I met the Christ.' Meagre and of an apostolic tastelessness, meals in the Carrière household were consumed in a silence occasionally broken by some beautiful thought uttered in a deep voice and with hieratic movements by the Master, in the small circle of light that streamed like a divine ray from the lampshade. In this pious penumbra Isadora met Mechnikov, Pasteur's successor and discoverer of the life-extending properties of yoghourt".

Savinio's remarkable early stories are anthologised in the *The Lives Of the Gods* [10] by the enterprising Atlas Press, publishers of the writings of Kurt Schwitters [11] Gunter Brus, [12] Adolf Wolfli [13].

"Perdita went to Doctor Sayas's and he asked her to undress behind a screen decorated with wading birds, standing on one leg at the edge of a river and others flying in spirals in the sunset-inflamed sky. Then he helped her get up on the articulated bed that occupied the middle of the office, on which fell the rays of three lamps grouped in the same crystal shell like three pearls fraternally united in one oyster. He inserted her feet into chrome metal stirrups that, like very thin arms flanked the bed. Finally he used his finger to search for the haemorrhoids of his beautiful new client - in a place where in living memory those, bloody engorgements had never flourished. "Doctor, you're making a mistake,"murmured Perdita, immobilised in the position of a bobsledding champion. "Don't worry, Madam," the doctor replied. "Just relax!"

On rare occasions major artbook publishers print artist's verbal creations where they accompany visuals. Being dead, or at least venerable, helps. Dover Books reprinted in facsimile Goya's pioneering volumes of sequential visual images with tangential, ironic caption texts: *Los Caprichos* [14] and *The Disasters of War* [15], and Max Ernst's surreal collage-novel *Une Semaine de Bonté* [16]. Thames and Hudson, one suspects, published *Roland Penrose's Scrapbook* [17] as much for its all-star cast of Penrose's friends, Picasso *et al*, as for the quality of his prose which culminates in an embarrassing eulogy of the art-appreciation skills of the Queen who knighted him. A far cry from the delicious anecdotage of Penrose's early bohemian life in Paris:

"At this time a familiar feature of the streets of Paris, easily detected by its pungent odour, was *la pissotière*. Its inner walls were plastered for intimate consideration, with dozens of advertisements of doctors, mostly quacks, who claimed they could prevent or cure all forms of venereal disease. Yanko confessed to me that he had at one time gone to see one of these 'specialists' in Montmartre who persuaded him that the most efficient means of avoiding such discomforts was circumcision and proceeded to operate on the young painter for a reasonable fee. This, however, obliged Yanko to return home on foot with nothing but a small parcel containing his foreskin in his pocket. The painful promenade continued across Paris as he turned over in his mind the appropriate way of disposing of this formerly precious part of his anatomy, until at last he reached the long dimly-lighted street that borders the Luxemburg gardens. There Yanko paused and delivered his burden to his most treasured friends, the cats."

Thames & Hudson also issued *A Humument* [18], Tom Phillips' labyrinthine dissection of a randomly chosen Victorian novel, but long after Ian Tyson's one-man Tetrad Press had bravely launched it. It also took twenty years gradual exposure in small press and postcard editions before major publishers realised the cult appeal of Glen Baxter's Boys Own drawings with absurd captions and, latterly, narrative texts in *His Life: The Years Of Struggle* [19] and *The Billiard Table Murders* [20]. Even so, Baxter's densely Rousellian prose pieces have appeared in their own right only in such small editions as The *Falls Tracer* [21] and *5x5* [22]

Limited funds inevitably restrict independent publishers' lists and rather than beat their heads against the walls of less sympathetic literary houses, artists may prefer self-publishing, as Ian Hamilton-Finlay did for many years with his Wild Hawthorn Press, and Wyndham Lewis with his review *Blast.*[23] Both artists used inventive typographic layouts which would not have been countenanced by conventional publishers. Another alternative is for the author to pay a jobbing publisher of what the book trade sneeringly calls'vanity editions' (without which we would not have the masterpieces of Lautréamont nor Firbank). In the introduction to *Abyss*, the novel she paid to have published, the painter Dorothea Tanning asks:

> "Where then is the vanity? Is this author eaten by ambition and vain glory? Hardly. (S)he comes on rather quietly, and his manuscript even more so. He will have no half-page ads in the book reviews. No TV appearances where he is obliged to sit in his place and look profound while answering banal or impertinent questions. No cross-country stumping to exhaust body and brain, no ladies clubs with ladies in them. Best of all there won't be any party at which the author is exhibited like a new breed of dog just officially recognised by the kennel club. There will be little or no distribution either, a not so happy feature. But it can't be helped."

Lack of distribution has always bedevilled those small presses who have, throughout the 20th Century, risked publishing innovative writing, especially by artists. Even in these days of 'financial realism' the labour of love continues. Hanuman Books, joint creation of the painter Francesco Clemente (illustrator of Savinio's The *Departure of the Argonaut* [25]) and ex-City Lights editor Raymond Foye, publish post-punk New York writers like Gary Indiana and Cookie Mueller alongside obscure tracts of the Beat Generation and the writings of artists such as Max Beckmann,[26] David Hockney, [27] and Henri Michaux.[28] The books' endearing Lilliputian format: 10x7cms, with candy-coloured coverings like Hindu prayer books, typeset in India by printers who speak no English, then hand-sewn by local fishermen, make them ideal for aphoristic mini-texts. Hanuman's most substantial volume, *The Collected Writings of Willem De Kooning,*[29] placated my impatience while waiting for a bus, and I

read Francis Picabia's *YesNo 30* in its entirety while boiling an egg.

As mainstream publishers become inflexible conglomerates, the survival of small, risky presses remains essential for artists' writings, to ensure the perpetuation of Picabia's dictum: "If you want to have your own ideas, change them often as you would a shirt."

Notes

1. Marc Chagall, *My Life* (Currently reprinting).
2. Roland Topor, *Stories and Drawings,* 1968.
3. Jean Cocteau, *Erotica* £22.50. *'Opium'* £7.95.
Le Livre Blanc, £6.95. *The Imposter,* £6.95.
4. Yoko Ono, *Grapefruit,* 1970.
5. Giorgio de Chirico, *Hebdomeros* (1924) 1970.
Giorgio de Chirico, *The Memoirs of Giorgio de Chirico* (1945) 1971: all Peter Owen, London, and all featured in *The Peter Owen Anthology,* Peter Owen, London, 1991. £16.50.
7. Alberto Savinio, *Capri* (1926), Marlboro Press, Vermont, 1989. £4.95
8. Alberto Savinio, *Speaking to Clio* (1939), Marlboro Press, Vermont, 1987. £6.50.
9. Alberto Savinio, *Operatic Lives* (1942), Marlboro Press, Vermont, 1988. £8.99.
10. Alberto Savinio, *The Lives of the Gods,* BCM Atlas Press, London WCIN5, 1991. £6.99.
11,12,13. Schwitters, Brus, Wolfli, all inc. *Black Letters,* Atlas Press, London, 1989. £6.99.
14. Francisco Goya, *Los Caprichos* (1799), Dover Publications, New York, 1967. £7.95.
15. Francisco Goya, *The Disasters of War* (1863). Dover Publications, New York, 1976. £7.95.
16. Max Ernst, *Une Semaine De Bonté* (1934). Dover Publications, New York, 1976. £7.95.
17. Roland Penrose, *Scrapbook,* Thames & Hudson, London 1981. £25.00.
18. Tom Phillips, *A Humument* (1970), Thames & Hudson, London, 1987. (Revised ed) £12.95.
19,20,21,22. Glen Baxter, *His Life, The Years of Struggle,* Thames & Hudson, London 1988, £6.50.
The Billiard table Murders, Bloomsbury, London, 1990. £13.50.
The Falls Tracer, Tetrad Press, London 1970,
5x5, Trigram Press, Hebden Bridge 1981.
23. Wyndham Lewis, *Blast 1* (1914), *Blast 2* (1915), Black Sparrow Press, Santa Barbara, 1981. £14.95.
24. Dorothea Tanning. *Abyss* (1947), Standard Editions, New York, 1977
25. Alberto Savinio, *'The Departure of the Argonaut'*(1918), Petersburg Press, London 1986
26,27,28,29,30. Max Beckmann, *On My Painting.* David Hockney, *Picasso* . Henri Michaux, *By Surprise.* Willem De Kooning, *Collected Writings.*
Francis Picabia, *YesNo*
All Hanuman Books, NewYork, 1990, at £2.99 each.

Marlboro Press and Hanuman Books are distributed in the UK by Turn Around, 27 Horsell Road, London N5 IXL

GLEN BAXTER

IAN BREAKWELL exhibits regularly at Anthony Reynolds Gallery, London and is also well known as a writer.
His illustrated fictions are published as *The Artist's Dream* by Serpent's tail, also *Seeing in the Dark: A Compendium of Cinema-going* (co-edited with Paul Hammond).
His *Diaries* were published by Pluto Press in 1986, serialised on Radio 3 in 1990, and in 29 Television programmes during the 1980s by Channel 4, who also broadcast his 5-Part autobiographical series *Public Face Private Eye* in 1991.

FRONTIER LORE

John Nicholson

How can we begin to understand the culture shock experienced by the natives and settlers of America? The contradiction of each value system is so immense, so vividly and harshly different, so total.

We did not ask you white men to come here. The Great Spirit gave us this country as a home. You had yours. We did not interfere with you. The Great Spirit gave us plenty of land to live on, and buffalo, deer, antelope and other game. But you have come here; you are taking my land from me; you are killing off our game, so it is hard for us to live. Now, you tell us to work for a living, but the Great Spirit did not make us to work, but to live by hunting. You white men can work if you want to. We do not interfere with you, and again you say, why do you not become civilised? We do not want your civilisation! We would live as our fathers did, and their fathers before them.
CRAZY HORSE, Lakota

So spoke the traditionalists. They saw no way to merge contradictions. Their way was under attack, threatened by death. They totally rejected an alien, murdering way which was death. On the one hand and on the other, the frontier is the meeting between alien values.

In recent years, around thirty, a start has been made by the descendants of the settlers to understand the culture their ancestors tried to eradicate. Very little attention has been paid by these same people to understanding and identifying their own culture. Consequently, we are expected to understand the victims in terms of how they responded, rather than look at why the acts were committed.

What kind of people did this; do this? What is the culture in whose name they act? To what extent is all Western Art tainted by inevitable association? Is the very concept of Art possible? Why fish up murexes?

I have deliberately created a large backcloth for the discussion, as I see Artists' Books, Small Press, Zine and other genres, as initiatives and attempts to create new and different forms. They are exploring the frontiers. Frontierspeople.

Although it is fashionable to patronise and hence package a myriad of initiatives under the term "Sixties" it is impossible to deny many serious (and frivolous) attempts were made to cast off the tainted inheritance. Scorn for the Sixties - a fake product and therefore an easy target - existed amongst rebels at the time. Long before Punk led the demolition squad which gave its bounty to Thatcher. Disarray? In the mid 70s? In the mid 80s? By the mid 90s? Which century? Such thinking has invisible assumptions: that disarray is a bad thing. That we must organise, march in a straight line, fight in a square. Of course similar criticisms were made to blame the natives - of America and elsewhere. They didn't use the methods of their exterminators. When the US army watched the destroyers of Custer's attempt at a surprise massacre they were afraid. Crazy Horse's cavalry-warriors wheeled out of the thousand strong 'village' in regimental formation.

But the natives remained true to their ways and within months it was all over bar the gatling guns. First the buffalo then the natives. By 1880, 600 miles of barbed wire came out of the factory every ten hours. The prairie died. In 1890 the frontier was declared to have vanished since it was everywhere.

To the worry about another frontier: between Art and the Book. I offer a way into these Badlands as a guide. I have been in and come back.

In 1960 I believed the Litmag was obsolete. With a few friends (didn't Lenin use this excuse?) in a small provincial town (who doesn't come from one?) we raised our flag of cultural regeneration. It was covered in the regulation spit and overwhelming indifference. It flew the regulation three and a half issues. Or rather it broke even the sequence of serial numbering.

The second issue destroyed the format established by issue one. It used technical tricks derived from Playboy magazine. Playboy had the money so it was the most exciting magazine technically of the time. We had a fold-out: a painting of a monk. A supplement was tucked inside a stapled fold in the back cover. It was a booklet of drawings by Aubrey Beardsley. Blissfully photo-copied with no regard for copyright. This was in 1963, before the famous V&A exhibition re-discovered him. (Indeed the pub in the village where the mag was printed was striving for a reputation as Patron of the Arts. The publican told us to pass on an invitation to our friend to have an exhibition in the pub.)

Issue three didn't go much further technically. Another fold-out (a map of concentration camps) and another supplement (a facsimile of Swift's *Modest Proposal*). This lost us supporters who thought we wrote it and meant it. Irony in 1963 was not a strong suite in English provincial towns. The fourth issue? Well we produced the supplement. It took all the poetry out of the magazine proper so it could breathe. But the magazine to go with the supplement never happened. So, for purists, *Skeletons* had six separate items.

If the numbering and format were awkward, the contents were worse. Political comment blurred with social invective and historical essay. What would be made in today's defiantly unshockable times of a series entitled The Prophylactic Conspiracy? Cartoons and jokes were replaced by inexplicable graphics such as a headless mannequin with a price tag. Prose could be a dialogue which was performable, or at least worth reading out to an audience. If there was any similarity in the mixture it could be the new satirical *Private Eye* . Far from the predictable mag we recognise thirty years later. In 1963/4 this was state of the art, indeed it was the cutting edge.

Others felt the same impatience, especially in America. We had comradely noises from other troublemakers such as a pair involved in *Fuck You,* a magazine of the arts which, set to music, became *The Fugs* who showed there was a niche which could be filled by the Mothers of Invention.

The frontiers were under heavy attack. The old pioneer *City Lights Journal,* could have been produced in letter press. Ginsberg was on the cover, miserable, crouched in a blanket on a rock. He was on the cover of EVO (East Village Other) two years later. Wild, crazy, beard flying, arms and legs waving, brain waving, HOWLING HAPPY ECSTATIC a wholly holy fool lost to civilisation-as-we-know-it.

What happened?

The new frontier meant more than a Kennedy slogan. It meant the exploration of *Inner space.* Drugs? Hippies? Psychedelic music? Alternative lifestyles? The underground? Dropping out - of what? The end of civilisation? A new consciousness, a new being? A war against frontiers.

The new newspapers of this community, like its social forms, blew away the frontiers. By comparison Litmags looked pathetic. They belonged to another age, to the old culture. In the non-newspapers minds swirled with pages of instant (semiotic subtextual) McLuhan messages. Collage and photomontage reinvented the visible world. Colours and dimensions blurred in and out, Escher-like. That was before you read a word. Writing? That tried to keep up. (I don't count the "New Journalism" which was a sales slogan.) I mean the unsung heroes beyond the forgotten ones. The next wave of rediscovered dead cult figures. Allen Katzman, Joel Kurtzman. Bob's trip to Allen's funeral.

New ways. Untainted ways. Not just aping, playing at niggers and injuns. But trying to face our culture.

They don't talk about that in the Sixties Revivals.

I made other expeditions. We tried to create a bookshop which ignored the frontiers. Non-literacy? A black-lite gallery in which suits and ties turned into stockings. Then another shop which destroyed even more frontiers. Was it a bookshop? Was it even a shop or was it a factory? A printworks? A publishing house? In one of my outpourings at the time I had called for a re-invention of the 18th century hot press, perfected in pamphlets and lampoons. Our venture also reflected another process at that time. The word was made paper in the bake-house, finished in the yard and sold in the front room.

Here is another prickly one — if you want it to be. Of course money was involved. No shop can be underground. But there is a wider picture. It all depends on how you look at it. If you like, it comes down to character, ethos, values - the frontier. What is for sale? The very notion of buying and selling is hazy. Trading was the old term. This need not involve money. The natives traded for *use* of the land, they did not sell it. How could they? That meant they *owned* it, which was clearly ridiculous. How could anybody *own* the Earth? "The love of possessions is a disease with them" warned Sitting Bull. "We have a disease which only gold can cure," Cortés told other natives. What value could be put on Paha Sapa? These hills were the centre of the world, its most sacred place. They were beyond value. But the settlers had other values. Custer told them there was gold in them thar hills. By 1879 one mine had been so successful it was floated on the New York stock exchange. Anybody who invested $10,000 saw their money grow. In ten years their stake would be worth $6 million. That is a good deal. Where are the descendants today?

When the settlers tried to buy *all* the Black Hills they had offered $5 million. In the end the natives lost the money and the hills. That is a poor deal. Where are the descendants today?

"The beootiful Indian countree that I lohuv" sang Calamity Jane, Doris Day, Doris Kappelhof. This sincere Jewish warbler, this cross-dressing Cowgirl, 'lost her heart' in them thar hills.

In 1939 the final head adorned Paha Sapa. The idols of the settlers gazed out from Mount Rushmore over the sacred sites. Sculpture? Art?

It depends on values.

"An idle and thriftless race of savages cannot be permitted to stand guard at the treasure vaults of the nation which hold our gold and silver...The prospector and miner may enter and by enriching himself enrich the nation and bless the world by the result of his toil." Contrast those values with Crazy Horse who despised the work ethic.

But that is an ancient crime. No it isn't. Paha Sapa

had more value. Even more 'precious' minerals were in its composition. In 1952 another 'rush' started - for Uranium. The military-industrial complex planned a vast energy empire in the West. Why even the dead-lands of the Navajo were perfect for bomb testing. The timber was chopped down, the surface was stripped off for coal, the water was diverted. "How can you sell the air?" asked a chief. I don't know tell me please.

Twenty years after *Skeletons* I was at it again. I went into the frontier with another expedition. It went beyond the non-newspaper and adapted the form to a non Lit-mag. The fold-out broadsheet used by *Rolling Stone* gave an enormous page! Here was scope for designers: graphics, strips, collage and montage or photos. What could be done with words and images if you treated the whole page as one? It offered a high level of possibility never available to solitary non-commercial creators. But there were other innovations on the business side. There was no publisher. No editor. We called it a "contributor" magazine and pastiched the idea in a pulp-style advert. Every contributor paid £60 for three giant pages. In return they received 200 copies (Work it out). The total print-run was huge, done on a newspaper press and so, dirt cheap. Apart from the guaranteed circulation, via contributors, we had another 4,000 copies to play with! So the only central organising was purely practical. It was a clearing house with no control. No imposition of taste, no interference in content. The contributors could make a profit if they wished and buy extra copies too. They sent to their own audiences and contacts which meant an otherwise unheard of publicity machine. Nobody lost. The dream was safe too. In retrospect I can see the shape emerging for a group of like-minded independent creators. A few years later I floated another version as *The Spectre*.

From our complex (was it a shop? a printworks? a culture centre?) we infected not only the town and area but other towns and cities. Years later we took some of the remnants to America and instead of the regulation spit and shit, we were covered in glory. It does affect you. We thought again. We saw the Past, Present and Future. We were told we had been a 'small press'. We were taken to a small press fair. We saw soul mates. Of course there were publications of the worthy cause sort but they were not valued above others. All lived together. We were thrilled by one-off editions which were like huge dolls houses, some three feet high and deep and across, with compartments and endless games. Were they books, art, toys, mad? Who cared? They lived. Or collections of rubbings of tombstones of bumper stickers. We were not alone.

A question of values. Small Press as we saw it was never a marketing device. Any more than Paha Sapa's incalculable worth was measurable in gold. If the dream places were desecrated then it would not be surprising if bad luck followed. But that is hardly worship based on a bargain. Of course it is possible to wreck the sacred places, just as it is easy to exploit a dream. They took the land and used it for their own material gain. Then they came back for the culture, the spirituality, the ethos - there is no word - to use it in the same way, for profit. As the land made them rich so the dream will make them civilised. Beware the miners, the loggers, the surface-strippers, the honest toilers.

Perched coolly on Glacier Point above a 3,254-foot drop, William H. Jackson composes a panorama of the Yosemite Valley.

Always pay close attention to your balance sheet

Nor were we alone in Britain. We knew a few others. We told them of the concept of small press. How it was neither antipathetic to money nor ruled by it. How money was not a value. How there were others who saw beyond money. The word spread. We learnt that in its short life the Small Press Group of Britain gave home to more than 900 members. It made some difference. Were all of them after gold?

Give me your hungry, tired and huddled masses and WATCH OUT!

An exhibition of Eclectic Formats from John Nicholson's private collection is scheduled for the National Small Press Centre in Spring 1995.

TRAVELS IN AMERICA 1994
DIARY NOTES
Natalie d'Arbeloff

Colorado Springs - When you first arrive, the light is so bright it penetrates all your defenses.The comfortable dusty veil you're used to in England is burnt away in a moment. You stand revealed with all your flaws and procrastinations and musty-maybes.

You take lots of very hot showers — zippy American showers that machine gun away all the English dust and suddenly you are a new you: a zappy, purposeful American you. Purposeful telephone calls of a marketing kind, rising at 6am and exercise before breakfast. Even a decision to go jogging every morning. But, the altitude soon puts a stop to all that: pounding heart, painful breathing after only a few steps, then splitting headaches and occasional nosebleeds. You're told it's altitude sickness (6000 ft. plus).

It snows. Then it melts almost immediately, the sun so fierce you can't stay out without layers of Highest Protection Factor cream on your skin. Everyone seems to be of a pleasing *café-crème* colour which you strive to achieve. After a morning's teaching all I want to do is sleep and sleep. The energy flame is cooling. "That's cool, totally!" These words constantly heard along with "Howzitgoin'" and "Howyadoin'" (no question mark as no reply is required). Impossible to meet one of the students' eyes without one of these phrases popping out of their mouths. As for "Haveaniceday," the smile that goes with it is sincere in this part of the country. People here have time for you, life is not rushed, the campus is beautiful, mountains and ever-blue skies and generous pay-cheques. Yet behind the students' laid back manner I detect a palpable anxiety. They attempt to drown it out with any kind of noise, the ears plugged into the ubiquitous Walkman, shutting out both the outer and inner voices. But the anxiety remains, that gigantic, collective American anxiety called AMIGOODENOUGH? The constant preoccupation of living up to some kind of image, whichever one is chosen as the current model.

By the end of my three-week teaching stint, the students have become my friends. We put on a truly impressive exhibiton of the work they've achieved during this 'Block'. We all congratulate each other. I feel proud of my sixteen pupils and stay up all night writing each of them an assessment of their work and attitudes. Professor Jim Tressel, who founded and runs the Press at Colorado College (and who invited me to come and teach this Block) gives me a tour of the excellent premises and equipment at the Press. And asks me to talk to his Book Art students and comment on their work, I also reminisce on old times at CC with friends I haven't seen since I last taught here in the mid-1970s when the extraordinary Bernard Arnest was head of the Art Deptartment. The painter/printmaker Mary Chenoweth (formerly Head of Printmaking at CC) arrives, all the way from Nebraska to see me.

So much of what one sees in America is easy to make fun of or to be aghast at. Yet it is on journeys to the States that I am overwhelmed by the kindness and trust of acquaintances and strangers alike. American friendliness is genuine and spontaneous. It does away with social protocol and simply throws open the door. And where one's work is concerned, the people one meets in the same field tend to be genuinely interested and helpful. If they can think of any information that may be useful to you, they will volunteer it without you having to ask.

However, it's impossible to be unaware of the dark side of this land. One Sunday morning, screeching police sirens bring me running to the window: An Apocalyptic procession of Hell's Angels is streaming down the avenue in the middle of the green and tidy campus, their long hair (grey on some) flying, black leather gleaming, eyes hidden behind dark shades, arms straight out on the high handlebars of their Harley Davidsons. Watchful police cars follow close behind. Later I learn that this is a memorial ride for one of their number who was shot in a bar last year.

A life class model tells me what it's like in the *louche* neighbourhood where she lives, where teenage gangs roam the streets in constant warfare with each other and innocent passers-by, and where it is not unusual for gang membership conditions to specify: Find a blonde. Rape her. Kill her. A beautiful and gutsy blonde student tells me she always carries a gun in her car and has taken lessons to become an expert shot. She also paints harrowing pictures of dying Aids victims, emaciated animals and figures in post-nuclear holocaust landscapes. And she wins prizes for larger-than-life ice sculptures in competitions up in the Colorado mountains.

On the way to Denver airport, the sky is huge, the light is astonishing; translucent, liquid. I wonder what this land looked like before it became scarred with motorways, motels, carparks and car-graveyards? Buildings look like they've just been dumped into this spectacular landscape, occupying it but ignoring it. I also wonder why the American Indians didn't build cities, temples, as they did in Central and South America? And another thing, why did English become the language of North America rather than any of the other tongues that the early settlers spoke.

Flying over Denver, garlands of light spread out like a jeweller's shop window. Emerald-green and coppery-gold, with accents of neon ruby-red. High above, the sky is dark, cold, indifferent to this glitter. It has seen the wild past, long before the land was tarted up in electronic finery.

Dallas: Policemen and other male Texans (and some female) walk with their arms held away from their sides, as if about to shoot from the hip. Do they copy this gait from Westerns or did the Movies imitate life? There's a new breed of really HUGE Americans; not in height but in girth. Not simply fat but mountainous, they have difficulty in moving these masses of wobbly flesh about. Has anyone done research on this phenomenon. Maybe it's the hormones in meat.

Hot and humid night in an airless room in a motel across the road from Bridwell Library at SMU. The door of my room is glass with a curtain to draw across (the night porter says that's so you can see anybody who might try to break in). See your attacker before the attack! Appointment at Bridwell in the morning to show my CREATION book to Isaac Gerwitz and Valerie Hotchkiss. Also meet Jan Sobota famed designer-binder and the library's resident binder/conservator.

Weekend in Austin, staying with friends Decherd Turner and his wife Margaret-Ann. Before his retirement, Decherd was director of the Humanities Research Center and before that, of the Bridwell Library. Known to everyone in the Book-Arts world he is a one-of-a-kind mixture of old-fashioned Southern courtesy, acerbic wit, sharp observation and sensitive scholarship. And he cooks a great barbeque!

I'm given the grand tour of the sights including Fiesta day in San Antonio, riverboat ride, and mind-boggling trip down memory lane at LBJ Ranch, Johnson City, where the ex-president is given almost saintly status, complete with statue seeming to bless the tourists as we make our reverential way around the vast estate in mini-bus with uniformed guide droning the rehearsed narration.

After Texas: New York, then it's a blur of non-stop motion, like a Futurist painting. Getting on and off trains, planes, coaches, taxis, dragging my suitcase-on-wheels behind me across vast University campuses, in and out of hallowed halls and hushed Libraries. Unwrapping and wrapping my bubble-cocooned bookworks in front of those with the power to say "yea" or "nay" or "maybe" or "we'll let you know". And whatever the result I always learn something and gather new impressions.

On a Greyhound Bus from Rochester to Ithaca I pass through a town called CANANDAIUGA . White wood frame houses with neat front yards in quiet almost deserted streets. Peaceful, folksy small-town America as portrayed in Jimmy Stewart and Gary Cooper films of the 40s and 50s. In Ithaca, my motel room has a fold down bed and mirror walls and lots of tiny free gifts: sewing kit, Woolite, moisturiser etc. I love the surprise of unfamiliar hotel rooms. From my fold-down bed I watch a startling TV documentary demonstrating that the pyramids in Egypt were in fact built by Martians who were trying to match monuments that had existed on Mars. This is proved by the photos taken by NASA on Mars which, when computer-enhanced show a face-like, pyramid-like structure. According to this theory, humanity existed first on Mars then came down to Earth. I'm ready to be convinced. Why not? anything is possible.

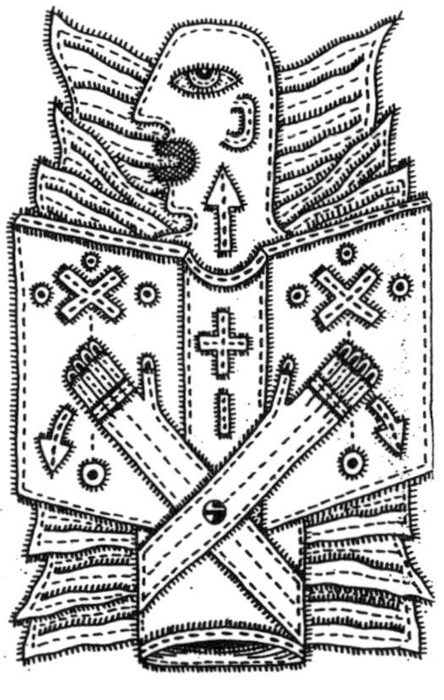

Image by Ian Pyper

BOOKING IN...

Concepts frame and envelop our thought and activity. If artists' books play within and without the conventions of books, then what is 'BOOK'? The meaning of all that surrounds us has evolved out of the past and I have recently become aware of the extent to which centuries old events are still having repercussions.

'BOOK' really began to have a wide social impact with the unique confluence of a number of developments in 15th Century Europe. The result, mass produced printed matter, gave the dominant system of western Europe a crucial advantage in the world... (See table one).

The book was one of the earliest capitalist commodities and a new medium of communication which catalysed and accelerated written forms of rational discourse and particularly science. Although manuscript books had existed before, emerging from the 1st to 3rd Centuries, it was the numbers and relative cheapness of the books produced by machine that gave this new medium special powers. History is always a fabrication based on the subjectivity of historians as much as on objective facts. This subjective interpretation is given expression through selection, exclusion and emphasis. (In the 'history of printing' timetable I am aware of attempting to counter the prevailing balance of history within the limits of my sources.)

The new medium was to provide a new weapon of power - propaganda - that was to change politics and the mechanics of oppression. It was also to gradually democratise the production and consumption of knowledge.

Both these effects changed language and encouraged the formation of modern national identities. Finally, and most importantly, the culture surrounding and mediated through the book was central to the formation of a new ruling class culture - epitomised by the 'man of letters'. (These effects which flow from and through the book are noted in the third table.)

So 'BOOK' is not a singular item but a bulk; a cultural mechanism; a reification of memory; a mechanism of thought; a weapon of power; a phenomenon; a tool for representation; a mirror of the mind; a repository of language; a filter or selector... The book produced a people of the book. Just as the page is made of wood pulped, so the people of the book are stately, magnificent and... wooden. Although a new electronic oralacy has swept the media tables, and books are now just one medium amongst many, the heritage of the book still commands the higher reaches of society.

Culture adapts successfully through the adult form of play known as art. BOOK needs jiggling free from the wooden aspects of its own past. This is why books and artists are a critical conjunction.

TABLE ONE ; CONDITIONS FOR THE BIRTH OF BOOK

0. **An alphabetic written language.** Use of our alphabet which first evolved from the old North Semitic script spread in roughly the same pattern as Christianity. More advanced but non-alphabetic civilisations could not develop an efficient and flexible printing industry.

1. **Codex format.** The technique of covering a wooden board in wax was the forerunner of the notepad. Parchment was then used in sheets, but, it seems, only as notebooks. Before Christ, serious publication was done in the form of the scroll, whilst commandments might be 'written in stone'. The scroll format was embedded in Judaistic ritual and it was only when the neonate Christians were looking for new forms in which to present their sacred texts that the codex format became a container of authority... a tabernacle of truth.

2. **Christendom.** A non-political religion which existed over a wide area of Europe and yet co-existed with a multiplicity of kingdoms and city states. The (written) word was sacred and monopolised by the monasteries. Throughout the See of Rome the language was Latin.

3. **Latin** and the Papal spectacle was the unifying force of Christendom. This international language also allowed access to the writings of the ancient Romans and Greeks. Many of these 'classics' seemed to have survived in Arabic culture and re-entered Europe with the Moorish invasion of Spain. The group of people who turned their attention to these texts rather than to medieval writings were to be called Humanists. The Humanists also read ancient sources in Greek and Hebrew.

4. **Technology.** The European Middle Ages produced machinery on a scale no civilisation had previously known. The history of technology in this period is suppressed in favour of romantic tales of chivalry. Is this because the knowledge was in oral circulation amongst a broad lower class of engineers? Later, in the Enlightenment, these and other knowledges of the people were corralled by the literary class. However it seems to have been these same oral knowledges that, in conjunction with ancient treatises by the great Greek natural philosophers such as Archimedes and Pythagoras, gave rise to what we know as Science.

5. **The establishment of secular universities** broke the monastic monopoly of knowledge and produced an increasing demand for books which the scriptoria could not keep pace with. The first university was in Bologna in the C11th; Paris university started in the C12th followed by Oxford. Cambridge began in 1209. Their main purpose was to prepare young men as cadres of the state, which they still do.

6. **Manuscript books and binderies.** In the 700 years between the fall of Rome and the 12th century knowledge was preserved in the monasteries by scribes writing in Latin on parchment or vellum, the skin of sheep and calves. Such books were both precious and rare. The new universities encouraged a secular bookbinding and stationers trade.

7. **Paper.** The production of paper took a thousand years to reach Europe from China along the ancient silk routes. By the time of Gutenberg, paper had already displaced animal skin as the preferred medium of scribes.

8. **Xylography.** The wood block print was already known and used for fabric printing, but it was only with the invention of paper that the wood block could produce a cheap, fine print picture. The early images were pictures of the saints or Blessed Virgin. This was the first time people could personally own a picture. It is difficult for us to imagine how potent such images must have been. This market, along with the new university-driven manuscript book production, primed the market for the flourishing of a printing industry.

9. **Metalworkers.** The die stamping and fine casting, and skill with amalgams, which was required to make movable type fonts, were skills that had been developed by goldsmiths and medallion makers.

10. **Money** and the development of trade slowly gives rise to a new class of financiers, merchants and manufacturers some of whom became as powerful as the traditional nobility. It was only these people who could finance paper production and associated printing workshops.

11. **Gutenberg,** an enigmatic figure of whom little is known, was central to the invention of movable type in Germany in 1450 before sinking back into obscurity. "...despite the many contemporary references to his role in the invention of printing, the name of Gutenberg does not appear in the imprint of any book." (Febvre & Martin 1990.)

12. **The Black Death 1347 - 1351.** Wiped out about one third of the population of Europe. This uniquely apocalyptic disaster must have caused a great trauma. It would have interrupted tradition and created spaces in which innovation could occur.

13. **Martin Luther's** protest against the See of Rome relied on printing in the vernacular to stimulate widespread popular debate. He called for a widespread reading of scripture which implied the democratisation of knowledge. Protestant culture promoted the word as superior to the image. A far reaching purification of the visual imagination was attempted.

A TIME TABLE FOR THE BOOK

3rd Century BC and earlier The first written records were kept on tablets of wood, sometimes loosely bound together. The most 'leaves' found so bound is ten. Other documents were written on papyrus scrolls and inscribed on walls.

84-86 AD The first unmistakable evidence for literary publication in codex form. (Roberts & Skeat, 1983)

105 AD The first writing paper was manufactured by Ts'ai Lun, chief eunuch to the Chinese Emperor. Knowledge of papermaking slowly spread out along the silk trade routes.

3rd Century The codex format on parchment had become ubiquitous and achieved parity with the roll.

868 The British Library keeps the oldest dated printed book, a long roll printed by xylographic process in China in the year 868.

1136 The Arabs brought papermaking to Spain establishing a mill at Cordoba which was a great centre of learning.

1147 Woodcuts were first used for printing the ornamental initial letters in manuscripts.

1275 A paper mill at Fabriano in Italy began making rag paper using a new stamping mill to pound the rags to pulp. This produced a fine paper surface suitable for the sharp quill pens of the medieval scribes. From this period, the rise of the Rag and Bone men.

1289 The first wood block printing of images on paper in Europe.

1300s A typical manuscript book size was 240mm x 160mm, with 150 leaves which would use about 10 skins each a half square metre.

1403 After a decree by King Sejong of Korea in 1403 a fount of 100,000 copper sorts was cast. During the century ten founts were cast, the first three preceding Gutenberg's more famous experiment. The Chinese may have been first to use movable type in the 11th Century.

1450 (approx.) The first European books, printed with movable metal type, were printed in Mainz, Germany. Commonly attributed to the press of Gutenberg although various people were working on this development simultaneously.

1461 The first illustrated book was a small collection of popular fables by Ulrich Boner, the Edelstein (Precious Stone).

1476 William Caxton set up a press in Westminster, London.

By 1480 there were presses operating in 110 towns throughout Western Europe.(Only four were in England, most being in northern Italy and Germany.)

1480s The Humanists, primarily Petrarch, introduced the Roman script, as a suitable style for their republishing of Latin classics. This differentiated the Humanists' texts from the Medieval scholars' use of Gothic. Gothic typefaces continued to be used for books published in the vernacular.

1490 The first globe. A dramatic new representation of the world constructed from calculation. From then on theoretical and abstract knowledge gained a new status over knowledge we affirm through our senses.

1498 Dürer produced a famous series of wood engravings illustrating the Apocalypse which were then published as books. Were these the first great artist's books?

Up to 1500 77% of the incunabula were in Latin with 5% in German, 5% in French, 7% in Italian and 1% in Flemish. Religious works comprised 45%, classical & literatures 30%, law 10%, scientific 10%.(Febvre & Martin.)

By this date 35,000 editions had been published which implied about 15 - 20 million copies. By now 236 towns had printing presses.

In 1517 Martin Luther nailed his theses to the chapel-door in Wittenberg. They were printed up in German translation, and 'within 15 days had been seen in every part of the country'. In the two decades 1520 - 1540 three times as many books were published in German as in the period 1500- 1520, an astonishing transformation to which Luther was absolutely central. (Benedict Anderson, 1983)

On 13th January 1528 "... the ultimate insult - someone posted a fake Papal Bull of Clement VII which permitted and enjoined the reading, re-reading and dissemination of Luther's works... All Germany caught fire. Pamphlets filled with the thunder of violent prose came out on all sides." (Febvre & Martin)

The portable format, which dates from **the first half of the 16th Century** becomes increasingly important. Wooden covers are replaced with cardboard made by gluing together scrap letters, business files, archives, proof sheets etc. Old book covers can provide a fascinating palimpsest. Books were still transported unbound in barrels.

16th Century In a period of ostentatious display, picture books spread across Germany and France. Both lavishly illustrated books and simple broadsheets with a single woodcut were produced in great numbers.

1539 Printers formed an unusual section of the working class as they had a high level of literary skill, often being required to read Greek and Latin. However, most printers were very poor, living in one room with their families, with little or no possessions. They formed brotherhoods or chapels and in 1539 all of the 1000 print workers in Lyons (and later Paris) are first recorded as going on strike after being threatened with wage cuts.

1543 Copernicus' *De revolutionibus orbium coelestium libri VI,* was published to scarce interest. Earlier book buyers had been more interested in the fabulous 'travels' of Sir John Mandeville than the geographical works of Marco Polo.(Only reprinted four times before 1500.)

Mid 1500s: the repression of heretical literature; "Most of those arrested or burnt were colporteurs, or occasionally small booksellers and boy assistants. Among the real victims of the repression cannot be found a single name belonging to one of the great printing dynasties that dominated the trade." (Febvre & Martin.)

1586 The first British 'Emblem book', Geoffrey Whitney's Choice of Emblemes. (Tessa Watt 1991)

By the end of the 16th century 200 million or more books had been printed. From this time the woodcut declined, although it was still much reused in the popular press...and copper began its long reign of more than 200 years. In the second half of the C16th 3000 distinct ballads had been published which amounts to some 3 million ballad sheets in circulation. This was the cheapest and most accessible form of print. Distributed through a diverse network of wayfarers, each sheet cost a penny or halfpenny (a labourer got 8 pence a day).

In 1500 churchmen had 24 major libraries and lawyers had only one. By **1600** lawyers have 71 whilst the churchmen have 21. This shows the shift of power from the church to secular authorities.

In 1615 "... the number of printers in London was fixed at 22 and the only authorised persons allowed to print outside of London were the university printers of Oxford and Cambridge." (Febvre & Martin)

1630s Education in reading reached a peak in England and gave birth to a new market for cheap books - chapbooks. By the late C17th the chapbook began to take over from the ballad as the commonest form of cheap print. These were quarto sized in a 24 page format with self cover, a format which lasted through the C18th and C19th. They included penny merriments, penny miscellanies and penny godlinesses, although by now they cost an average of 2 pence. (Tessa Watt 1991)

It was not until the **C18th** that a definite standard of measurement of type, the point system was agreed. The point was one 144th of the size of the King's foot.

It is only in **1778** that we have the principle of authors' perpetual copyright established, with the bookseller or publisher holding a temporary copyright. The private patron or state pension was exchanged for a share in profits.

1791 *The Rights of Man* written by Thomas Paine, a self publisher and friend of the greatest artist book maker, William Blake.

1814 John Walker, director of The Times, introduced a new mechanised press. This can be seen as the birth of mass media, mass literacy and modern hegemonic mediation of power.

20th Century Invention of oral mass media (film, radio and TV). The return of the image and its challenge to the dominance of literary representations.

1950s Penguin cheap paperback editions. Working class writers invade the new oral media, especially TV. Reversal of the trickle down effect with a remarkable (sub) cultural renaissance from the working class.

1960s Collection of small presses and little magazines started at University College of London in 1965 (Soar and Miller 1994)

1980s The book is ubiquitous, ordinary people have libraries, but the rise of the personal computer challenges the sacred status of the book as a receptacle of knowledge. Book metaphors and conventions are used in this transformation just as the first fonts imitated the black letter script of the scribes...

1990s James Kelman wins the Booker Prize.

THIRD TABLE ; WHAT IS BOOK CULTURE?

The book was the primary capitalist commodity. A newly commissioned book was, and still is, considerably more costly to produce than a reprint. The rich people's market directed the content of new publications. Reprints, as cheaper editions, would be available to a wider market. So it was that the choice of content, including language and spelling, of the upper classes would later be passed down to poorer classes. This is part of what is known as the trickle down effect. There were broadsides (printed folksong); and then chapbooks (including myths and folktales); and also pamphlets (polemics on current affairs which were the precursors of cheap newspapers); all of which were a kind of printed oral culture but these productions were dominated by 'real books'. This image was partly based on the superior skill of the professional writer but also on the selectivity of rich publishing dynasties working closely with the academies like Oxford or the Sorbonne and the arbitrarily constructed superiority of certain formats and styles. Standards were guided intuitively by 'Good Taste' (simply the conventions and values of owning class culture which are mythologised as 'civilised' or inately superior).

This market relied on the exploitation of the print and paper workers. As some of the earliest workers in a mass production process they worked twelve to fifteen hours per day for subsistence wages. From the C16th they organised strikes and wrote and printed manifestos of worker liberation. Printers were often to be found in the forefront of progressive working class movements also dealing in radical, heretical and banned books.

The militancy of the early print workers caused a reaction from the print bosses who closed ranks and instituted a series of repressive state ordinances.

The tendency of commodity markets to find the maximum number of buyers, led to the evolution of vernacular print languages influenced by the dialect of the main print market. In Britain this was a compromise of Oxford and London. With the authority of print and the mechanism of mass production, this tends to flatten the diversity of spoken vernaculars and to produce a nationally dominant and standardised form. These have been held up as 'correct' in an absolute way rather than as simply reflecting the evolution of conventions within a context of market forces and class domination. This common language of print helped to produce the modern national consciousness.

Our idea of what constitutes knowledge has been filtered and validated through the literary and academic elites. This concept of 'knowledge' is so much a normal part of our intellectual backdrop that it is invisible and also mythologised as neutral, rational, objective, scientific, rather than a very particular and biased view of the world directed by the consciousness of upper middle class imperialists. The book, particularly as organised in libraries, reifies this world view with its emphases (in terms of quantity and quality) and perhaps more important, its blanks and silences. Public libraries obviously do not leave empty shelves for the blanks of History and Knowledge so they are not obvious! The silences too, are by definition, not easy to hear... until they get filled from alternative sources or our own experience. Sources which are rarely privileged with the same authority.

The categories and subject headings into which our knowledge of the world is ordered and divided is reified by books. This is reflected in wider educational curricula. The whole set is taken for granted by most people as a neutral representation of the world rather than a partial selection produced by the self-interest of a capitalist class. Printed matter has made oppression possible with less overt violence through a widespread belief in cultural values of oppression.

The book as commodity produced the possibility of propaganda. The bourgeoisie evolved with this powerful tool at the heart of their culture. We have already mentioned the astounding use of the press by Martin Luther who went on to standardise and reform Germanic languages...

Our current idea of the level of cruelty of the Spanish Inquisition is a result of early printed propaganda by anti-Spanish forces in the rest of Europe. It is only recently that a computer analysis of the records of the Inquisition has revealed it to have been, if anything, less cruel than other horrific practices of repression which prevailed throughout the rest of Europe. (This is not to forgive or veil its anti-semitism.) The aristocracy thought the use of propaganda most unchivalrous. Honour demanded that they assert power and wage war with the sword, not with a pen.

Because of the class connotations of the literary we find that the culture of the true middle classes is underpinned with terror and pretence. Knowledge is not an open field in which everyone is invited to wander and delight. It is a hierarchical canon with a range of key facts which one must know or give the impression of knowing. Admitting ignorance is the terror behind every middle class conversation. As published knowledge becomes ever more diverse and bounteous this game becomes increasingly absurd. Still the canon is defended with a fanatical zeal which is only equalled by the perverse passion for correct spelling and pronunciation.

The authority of the book can affect the way people think. The mentality of letters is dominated by words. Sentences and phrases constantly fill the mind. Impressions and sensations are instantly translated into verbal forms. The linearity of the text also suggests the primacy of a linear form of argument. This is a very simplified model of how the human brain actually makes judgements.

Oral thinking may use similar concepts but they are more sensate in their formation, articulation and synthesis. Literary thinking becomes more individualised, a singular rather than a social discourse. It is difficult for the literary person to imagine a crowd thinking...

History is constructed from literary documents. It becomes the history of those who wrote documents or through the eyes of those who wrote documents.

We are the total of our past, the sum representation of our memories. The social perception of our past also relies on representation. Our sense of ourselves in the social present is therefore based on the representations which were largely written by a class which dominated and exploited the majority of people. Our sense of ourselves as human beings is therefore often seen through the distorting glass of oppressor consciousness.

Success within book culture requires sitting still. A physical and emotional passivity is required. This fits in very well with the values of dominance which prizes the abstract and mental and devalues the physical, sensory, sexual, emotional. The book skills and oppressor conditioning get conflated and confused. Academic style idealises the non-emotional, the cool, the cerebral. It cannot easily accommodate commitment or outrage. Conceptual agility and abstract gymnastics are valued above useful clarity and wider accessibility.

SO WHY IS 'BOOK' AND 'ARTIST' AN IMPORTANT CONJUNCTION?

Any society or culture must play to adapt successfully. The market is inevitably led by short term profit - however much eco-style corporate image this is given. We cannot rely on the market to innovate with our long-term well-being in mind.

We have a world stage built of books. On this stage the players are watching interactive TV. The world has changed. We are in a position to reassemble our relation to each other and the world but we cannot do this with the woodpulp mind-sets of a literary heritage.

If we ignore this heritage in the rush to speed down the information superhighways we will only take the dead wood with us into hyped-up-space. We need to play out the codex legacy with an awareness of how we have internalised its perspectives. The key players may be book artists.

Stefan Szczelkun 1995.

BIBLIOGRAPHY

Lucien Febvre and Henri-Jean Martin, *The Coming of the Book, The Impact of Printing 1450 - 1800,* translated by David Gerard. 1976 NLB, (originally Paris 1958), most recently Verso 1993.

Tessa Watt, *Cheap Print and Popular Piety 1550 - 1640,* Cambridge U.P. 1991.

Benedict Anderson, *Imagined Communities, Reflections on the origin and spread of nationalism,* 1991 (Rev. and extended ed.) Verso

C.H.Roberts and T.C.Skeats,*The Birth of the Codex,* Oxford U.P. 1983.

ON THE LIBRARY'S RUINS

Simon Ford

When Norwich Public Library burnt down in 1994, the deeply felt public expression of sadness and loss was both predictable and justifiable. However, in some quarters news of the disaster was not met with horrified disbelief but with unrestrained joy: biblioclasts everywhere were celebrating the book carnage. Michael Foucault has written: "Rhetoric is a means of momentarily postponing the burning of libraries". This statement follows on from the argument used by fundamentalist book burners; that either everything a book can say has already been said in the holy book and is therefore superfluous and must be burnt, or it contradicts the holy word and is therefore evil and must also be burnt. Book burning is still probably the most emotive symbol of barbarism to the civilised. So, what is postponing the burning of libraries today? Could it be that there are now much less spectacular ways to destroy libraries?

The biggest threat to the library are the new forms of electronic communication, such as multimedia. With competition from the electronic book, the codex book is slowly dying, and the end of the library, as we know it, is not far behind. The library is not disappearing with a bang but with a whimper. The call to destroy libraries is not new, although it is often observed that the leader of the Italian Futurists, Marinetti, in the *Foundation and Manifesto of Futurism* (1908), called on his followers to destroy the museums and galleries, he also called on them to set fire to the library shelves. This call to burn libraries was repeated by the artist Ben Vautier, and Art Total in 1964 when they also considerately provided the matches. And just in case you thought artists had grown out of this kind of thing I recently received a package through the post, from a group calling themselves, Field Study, consisting of some 'burnt offerings' and an incendiary device with the instructions to 'Burn down the libraries of Alexandria!' Another example of artists' hostility to books and libraries is provided by the Situationists. Guy Debord and Asger Jorn, who in 1959 covered their book *Memories* in sandpaper.

Libraries, rather than being disheartened by such attacks, are instead returning to some of their earlier museum-like functions to collect, preserve and interpret the art and history of the book. The library is to books what the museum is to art. This formulation is complemented by the genre of artists' books, a genre that brings together two of the avant garde's most sickly and hated enemies" the 'art lover' and the 'book lover'. With artists' books the book moves from being a secondary source for another subject to becoming a primary source for its own subject.

Increasingly databases and CDs are freeing books from their duties as information providers. Reference works such as encyclopedias are now much more flexible and useful when published digitally. This does not mean that books will disappear entirely, just as painting did not disappear with the invention of photography, books are rather in the process of changing their social function and entering the realm of art and craft. And what happens to books when they are produced as works of art? They gain new status and enter realms of privilege and power that promote particular values and meaning.

The library and the art library in particular, that collects the material, now promotes a variety of values which range from the aesthetic to the religious and from the entertaining to the moral. This widening of the library's function makes the library more visible as its activities become more problematic. For example , what could possess a library to acquire, catalogue and conserve a screwed up piece of A4 paper? Last year the National Art Gallery acquired just such a piece called *Work no. 88* by the British artist Martin Creed. It was produced as part of the Imprints '93 series of artists' books published by Matthew Higgs and I wrote the justification for its acquisition.

Sometime during the 1960s artists' books began to move away from the exclusive and luxurious *livre d'artiste* tradition. The book now became an 'alternative space' and a cheap and democratic means to reach larger audiences. Just as some artists such as Hans Haacke and Daniel Buren began to critique the institutions of art, so did artists such as

LIBRARY

THINK TANK

Ed Ruscha with his much cited *Twenty-six Gasoline Stations* (1962), begin to cause headaches for librarians. How do you classify such a work? Is it a book about commercial architecture? Is it about automobile culture? Or is it even art? Of course it was all these things and more. Artists' books began to be produced that questioned how libraries ordered their collections.

Artists were not alone in their battle against antiquated library procedures. Since the sixties librarians and their libraries have changed considerably along with the materials they collect. The most significant of these changes is the attack on the concept of a received canon. Other changes include the widening of what constitutes legitimate areas of study, in particular with cultural studies and design history and, in connection with this, the incredible increase in the amount of theory produced. Theory is now central to debates about contemporary art and culture, both high and low, and challenges the 'picture book' as the major source of inspiration for the contemporary artist. Artworks often come with bibliographies attached, whereas some artworks, such as Anselm Kiefer's *Book Wall* are actually libraries.

It becomes difficult to know what to place in the library because now everything is of potential value. Art's autonomy and distinction is eroded by cultural studies where Mills and Boon texts are as potentially worthy of study as the latest Rosalind Krauss blockbuster. The library is now the place where there are no agreed standards or natural classification systems for organising knowledge. It is a place where it is increasingly difficult to identify and separate categories and where order often gives way to disorder. The library, like many of its users, is becoming decentred and confused. This has meant that librarians have had to revise their acquisition policies and review how their material is organised, catalogued and accessed; in other words the basic assumptions of library practice are being re-examined. The library's relationship to power is at the core of this re-examination.

In Umberto Eco's *The name of the rose* one of the characters observes that "The library is testimony to truth and error". A celebrated example is the so-called 'wicked bible' of 1631, complete with what is probably one of the most disastrous typographic errors in book production history, where one of the Ten Commandments incorrectly read "Thou shalt commit adultery". In the Middle Ages the library was a critical facet of the Church's power and since the Renaissance magnificent library buildings have been testimony to how political power values 'the library' as a spectacle of its own glory. The great libraries make the concept of 'civilisation' and 'culture' real and tangible. They are the material evidence of the great stories of progress, embodying and making perceptible the Enlightenment project and its values. The library is thus a legitimating institution that has a symbolic and functional power equivalent to that of the museum and gallery.

Books acquire legitimation firstly by their acquisition and then secondly, according to the place they are given in the library's classification systems. The library substantiates, services, and maintains a real, if not always clear, order of value, which reinforces distinctions made by the dominant culture. This is most obvious in the designation of some material for inclusion in the so-called 'Special Collections' . As information becomes commodified 'knowledge is power' has also become 'knowledge is money'. The library is totally caught up in this process and many of its most cherished human pretensions are threatened by it. This is particularly clear in the growing restrictive practices surrounding copyright.

Knowledge is structured, and cultural value represented, through the ideology of library rules and organisation. Probably the most famous instance of an artist breaking library rules was supplied by the biblioclast's biblioclast, John Latham, and his non-return of a library book to St. Martin's School of Art. In 1966 he and his friends chewed and then spat out, into small jars, pages from the library's copy of Clement Greenberg's *Art and Culture.* These remains were then fermented and distilled. When he presented this liquid 'essence' of the book back to the librarian, Latham was dismissed from his teaching post. He eventually sold the masticated remains of the book, neatly packaged in a briefcase with documentation, to the Museum of Modern Art, N.Y.

The library is a place of many rules, standards and procedures. It is not surprising that librarians are often likened to jailers and books likened to prisoners. The artists Langlands and Bell also made the connection between penal servitude and libraries quite explicit in three books in 1987 which explored the similarities between the floor plan of the British Museum, the round reading room of the library, and the panopticon prison of Jeremy Bentham (plans of which were published in 1791, the same year as the first standardised cataloguing rules). The birth of the modern library is closely connected with the modern period of discipline and punishment.

How knowledge is organised (classification) effects how we see the world. The Argentinian librarian and writer Jorge Borges describes a classification scheme from a 'certain Chinese encyclopedia' in which it is written that "animals are divided into (a) belonging to the Emperor, (b) embalmed, (c) tame, (d) sucking pigs.." and so on. This passage illustrates how classification can be arbitrary as well as dependent upon ideological, historical and utilitarian concerns. One person's order is another person's disorder, and other 'systems of thought' reveal the strangeness of our own. Classification and forms of cataloguing do not just aid the retrieval of relevant material they impose limits upon what is accessible and therefore what is thinkable. Books were not always catalogued and classified as they are today. There are many curious examples of types of knowledge and organization such as the catalogue of a library in the eighth century A.D. which was composed in the form of a prayer, and the monastic library catalogue of York written in verse. The first systematically arranged catalogues did not begin to appear until the thirteenth and fourteenth century. Alphabetical author catalogues did not become customary until the sixteenth century. Still, the introduction of the computer catalogue has lead to much anguished debate over the merits of card catalogues. Interestingly, rather than be thrown away or archived, the card catalogue at the Widener Library in America has been given to a conceptual artist, Tom Johnson. This is an excellent example of how superseded technologies come to be reclassified as art.

Classification is now following developments in the field of computer science where flexibility is the 'order' of the day, superseding the idea of pigeon-holing knowledge in precise positions on a fixed two-dimensional diagrammatic space. The disorder of systems such as multimedia are replacing the linear order of the card catalogue. Current trends in cataloguing are bringing ever closer the object described and its description, the book and its virtual twin are becoming one. In the case of artists' books, cataloguers are still concerned with the conversion of visual information into textual information through controlled terminology. In the near future, library catalogues will become multimedia based and carry visual information as well as bibliographical information. The advantages of multimedia have been well rehearsed elsewhere. Its main effect is that it supersedes forms of communication that rely on ideas such as centre and margin and replaces them with concepts of paths, webs and networks. Multimedia does away with fixed sequence, with beginnings and endings, and destroys the idea of a text's singularity and unity. One of the most radical aspects of multimedia is its utilisation of images. The mind can assimilate images far easier than text: this is why most of the traditional 'art of memory' systems involve associating concepts with images.

It is unlikely that every text or book ever written will be digitally encoded and retrievable, and even more unlikely that a future internet will become a "Superpanopticon' where every exchange is supervised and where dangerous, radical and potentially threatening material has restricted access. Along with improvements in the dissemination of information there may come improvements in the technologies of control, the internet could take on the worst habits of the library. For multimedia to fulfil its potential it must be accessible to a wide range of users. The library may have some role to play as a bureau for providing searches for those without the machines or for those who do not have the necessary search skills. But there are few reasons for these terminals to be housed alongside books or in places described as libraries. Databases do not require the vast intermediary technology that books require; such as catalogues, reading rooms, and enquiry staff. It is more likely that the intermediary will be internalised, part of the computer/user interface. The information not on the internet, because it is obscure, of minority interest or in a non-western language, may become further marginalised, and practically invisible. The library can be a place to collect and conserve such important material.

Another intriguing thought is that as each generation becomes more and more orientated to the new media, books are becoming increasingly difficult to read and, as with manuscripts, interpretation of this material will become the domain of specialists, be they either curators or librarians. There is now a generation for which the book has not been the primary source of culturalisation. Students are more likely to watch TV and video and listen to CDs than read books. What the new technologies have done is 'make strange' the book and enable us to see it afresh, as a historically specific and soon to be superseded form of information technology. Artists' books such as Colin Hall's *Book in a jar* (a decomposing book submerged in a jar full of murky liquid) literally deconstruct themselves, becoming impossible to read in any conventional sense. This type of book has become so prevalent in the National Art Library that we have adopted the genre 'unopenable books' to group them.

The origin of the library was closely linked to the origin of the museum. Both were connected to the wider context of the collecting and organisations of curiosities, artworks, and natural history specimens. It was not until books were mass produced that they became increasingly subject to specialisation and isolation in dedicated spaces. This process was accelerated at the turn of this century with the gradual introduction of mass literacy. The roots of the modern library and the modern museum are both linked to the belief in the power of books and artworks to act as deterrents from vice. Do-gooders and reformers were the main force behind the establishment of both public libraries and museums. But this role for the library is fast disappearing as the book loses many of its older functions and becomes recognisable as an historical artefact. Librarianship increasingly resembles curating with the collection, preservation and interpretation of books as its main functions. This happens at a time when museums are becoming more like libraries. Museum collections, in the form of images and textual information, are being put onto CDs, in particular by Bill Gates and his Continuum project. The 'museum without walls' is close to becoming the 'museum without objects', a museum that can be networked in the same way as other digitally encoded data. The discourse of librarianship is partially surviving this change-over period because computers are often described using the terminology and imagery associated with libraries, such as Microsoft's *Bookshelf* software. The library is used to represent the invisible processes that take place inside computers because libraries are familiar and can represent something that is unfamiliar, and essentially unrepresentable. This, however, is a temporary feature, similar to when the first motor cars were made to resemble horse carriages.

Borges in *The Library of Babel* wrote that humans would soon be extinguished: "but the Library will endure: illuminated, solitary, infinite, perfectly motionless, equipped with precious volumes, useless, incorruptible, secret." Although I think humans will be around for quite a while, the library is indeed returning to its hermetic roots described by Borges. The new technologies effect the very structure of comunication and free the library from many of its eighteenth century conventions. The library becomes dangerous, a site for transgression and subversion, where world views are continually thrown upside down. I will end by mentioning what is probably the most hazardous type of book a library can buy. A book that could mean the end for both the book lover and the art lover, and the place where they carry out their dangerous liaisons, the library. A recent acquisition by the National Art Library is *Book Bomb* (1994), a replica pipe-bomb hidden in a book by the American artist Gregory Green. This type of book could be one way of ensuring the library ends with a bang rather than a whimper.

February 1995.

Simon Ford is a cataloguer at the National Art Library, Victoria and Albert Museum, London. A version of this essay was given as a lecture at the 'Ideas in progress' seminar at Middlesex University on 17 January 1995.

AN INTERVIEW WITH JOHN LATHAM

"A Man staring out of the back of his head"

Stephen Mepsted

When did you get interested in the book form? What were your thoughts as these works first came about ?

I wanted a surface that was distinct from the wave form I was using (corrugated asbestos sheeting). I had been making paintings using the atomised mark which is non-linear - the book has flat marks in a linear form but that was not what I saw. I'd gone to the wave form from the flat surface because I'd worked out what the painting would do and had come to the end of that: All the kinds of abstract and figurative things that were around and are still around and still working. But I realised that one can't go on doing the same thing - well, I can't go on, I can't make it an expertise. The wave form was something simple that wasn't flat, so I looked around for a flat thing to make sense of what the surface could be meaning, and I just happened to look at a book. It happened to be the right proportion to use with my piece of corrugated asbestos. It seemed a dodgy thing to do but I put the book into it: I cut the book. It was because the form was dominating the habit, the referential habit of the word, the form became more important than these unread volumes that are repeated. So, the book became as important as any other material, it *is* a material.

What was the response to that? Did people see this as an act of irreverence?

Well, I first took three pieces on asbestos to Victor Musgrave, a dealer. He thought they were very interesting - and I thought so too! So I left these pieces with him and after three or four weeks I hadn't heard from him so I went along and asked him about the response. "Well," he said, coming down the stairs and doing his trousers up, "I have to tell you that the response has been extremely bad, people have called them Victorian tombstones, it's really not helpful". I was a bit depressed with that but I wouldn't take it too seriously and in the course of the next month or two I went back onto the flat surface with the *Little Bagatelle Board* that is now in the Tate collection. The bagatelle board suggested to me that there was an equivalent for the El Greco painting with the rounded top in the Toledo Church. I was also excited by the idea of a whole equivalent for Heaven and Earth, this temporal/atemporal idea. What El Greco had done was to put them together; the omnipresent with the present and it's all personalised. I wanted to make a visual metaphor for it so I got all these odds and ends: a whisky bottle, a sponge, a key, very useful, practically everything I saw. I rolled up corrugated cardboard and made them into personages. And then of course there's a burial scene, and a fire god! God then turns up as a flint stone!

So, you were dealing with a great deal of objects and their physicality. From what you are describing, it sounds like you were having a great time!

Oh, it was wonderful! I really enjoyed it. I was sort of feverishly involved with the whole thing and it was a very formative time. It was around this time that I came across a Polyfilla type plaster - but before its existence, there was a kind of gesso stuff that you got from DIY shops. That stuff, when it adheres to a book, doesn't contract, it was stable as opposed to plaster which does contract. The use of this stuff was enormously helpful, just as recently silicone gel allows the joining of a book to glass. That releases a load of things that can be done that couldn't be done otherwise.

History is important of course, we need to latch on to what has gone on before, but we don't want to be believing that it's what we've got to do. The main thing we get from museums is a development so you know who you're talking to and you can develop further and be sensible in a new way.

When did you begin to use the book as a metaphor for the intuitive 'incidental' person and what meanings did you glean from the physicality of the book?

The book is like a vocabulary in form, you can face it in any direction, you can relate the different forms, you can bend the covers back and it becomes constellatory and it's so malleable - but it's also restricted. It is a good equivalent for the figure. We don't have to imitate the figure any more, there is an abstract geometry which will work instead of that, and I used it of course as parts of the person and this grew. The pieces with plaster spilling out are a good equivalent for the sort of person that I feel I am, and recognise as being in a muddle sometimes. Also there are certain genetic orderings that everybody has; the genetic character is the same and they are in fact very similar to what we call a genetic code, they run an event from start to finish and it does reach an end. The whole geometry of the book works out that body event is insistently recurrent. In literature this tends to be related in very succinct ways. There is an example in Dostoyevsky with the three brothers, the naïve desperado type, the intellectual who is insufferably clever and knows it and has to do away with himself because he can't stand it either! And his younger brother who is intuitive and doesn't care too much about either of the others as long as he can get along with his friends. Well, that third brother, that is the one that would work, that is the attitude that will get people through unknown circumstances whereas the others are tied to a genetic code and a library of intellectual concepts and theorems to determine behaviour. Another example is Joyce - *Finnigan's Wake,* which people say is too difficult but of course you don't pick it up and read it as a book! Well the reference to the skeleton key to *Finnigan's Wake* was something I found extremely affirming. The other thing was hearing Joyce reading a passage from the book where you can hear him going through the poetry of the sound he was thinking when he was writing. So, all of it is really an action which is heard as much as read and it is of course, poetry.

I remember the Lisson gallery show you had about two years ago now. You showed glass wedge shapes with books and other objects arranged within and around them. I seem to remember the books were all similar - one was on pregnancy I think.

That's right. They were remaindered books found in a warehouse and I bought them. I was asked to do a multiple by some friends and I needed the material. I looked around the warehouse and I found a book called 'Vanished Species' and another called 'The Pregnancy Survival Manual' and these together had everything I needed!

One of the things I noticed about the forms at the Lisson show was that books and objects were placed at different heights along the glass wedges. No two levels were the same and I couldn't find an overall pattern. Were you saying something there about levels of human intuition?

Yes, the final one of the series has a plane that is high up on the glass and one that is entirely free of the glass - it's an innocent state; all these levels have no faults attached to them, they are just states. That is why I don't get involved with politics in my work. There is so much one can argue for. There is no fault in the level; there is a relationship between the temporal and the atemporal and once you have the relation of the body event passing through you are in some sense valid, these things are validated in all the states.

I think that it is important for the work that you say that these states are 'innocent'

I'm glad that comes across. I have been hassled by so many people who say that I am not serious in the political area.

I would have thought that the work that you have done with the old APG (Artist Placement Group) and now, to a certain extent with O+I (Organisation and Imagination) should counter any argument that you do not get involved on a political level.

Well, it's got to have a political dimension, but it is the artist who is the left-winger or the radical, and obviously and necessarily so.

So, you do practice what you preach. I think you are saying that, in an ironic way, one needs an innocent, almost naïve and intuitive approach to our experiences in order to be truly radical.

Well the innocent, incidental person approach does dispense with a lot of noise and hot air and fruitless discussion about a matter that doesn't have a solution. Until we've actually seen the structure of the thing in which we are floating or making our way, the arguments are all spurious and quite redundant.

kArtBoo

Since the advent of printing, it's been downhill all the way for the poor old book. Prior to that the scratchy pens of nuns and monks scratched away day after day, mainly reproducing liturgical texts. On saint's days the production was generally for within the religious order. On ordinary days they would knock out slim volumes for local aristocrats, and in later years even an odd tome for a bourgeois.

The advent of printing changed all that. In England, the evil Mistery of Mercers had one of their number learn to print on the continent. This man, William Caxton, was soon whacking out a book on chess not a stone's throw from St. Paul's Cathedral. The book had entered the realm of COMMODITY PRODUCTION — in fact the book became one of the first commodities, in that as a finished product, each individual copy was ideally identical to every other. The text was to be reproduced uncorrupted.

And now, hundreds of years later when the role of the book is being undermined by electronic media, the book is being abased to a level even below that of the simple commodity — the book is being turned into *Art.*

This invasion of art into an area previously quite healthy and dominated by horny-handed sons of toil is to be welcomed. The Printers for many years had contested bourgeois control of the book industry. General stroppiness and intransigence held back the introduction of the new technology, only to be swept away through such strikes as that at Warrington and finally Wapping. Since these defeats it has been necessary for the proletariat to regard book production in a different light. The sooner the book can be reduced to the level of *Object d'Art* the better.

However this invasion has been proceeding slowly all to slowly. As most artists quite rightly have the last traces of originality beaten out of them at art school, this makes it hard for them to keep up with the tasks imposed upon them by the force of history. This lack of originality is of course superlative in its general demoralising effect upon society as a whole, and the spectacle of the self-obsessed sons and daughters of the middle-classes being trained to wallow in their emotional shallowness gives everybody a good laugh. However when it comes to reducing the book to mere vanity, these excellent qualities are not quite what is required.

With this in mind, *Unpopular Books* set out to up the stakes using techniques such as abstract thought, garnered from such diverse disciplines as mathematics and philosophy. While much Book Art was immersed in creating a parody of the old artisanal skills of the printer, *Unpopular Books* has always prided itself on the shoddiness of the finished product — e.g in our 1987 production of *What is Situationism* by Jean Barrot, we pasted in a piece of ordinary typing in the middle of the beautiful typesetting. Also a lot of the text was unevenly exposed during platemaking.

It was also with this volume that we started experimenting with the alternative *Standard* and *De Luxe* editions (achieved by adding a silver sheen to the front cover and adding an extra 5p to the price). Aside from increasing the sales to the Situationist Train Spotting fraternity, it was possible to baffle shop keepers and customers with a virtually meaningless choice.

However in 1993 we decided it was necessary to open up a new front with the libraries. There are laws in place requiring publishers to lodge copies of their books at six libraries in the British Isles. Administratively this has been tied up with the issuing of ISBN numbers. According to the regulations, books with different covers need different ISBN numbers. However this does not prevent a publisher providing some sort of throw away wrapper, which is not an intrinsic part of the publication. This may appear to all intents and purposes to be a cover, although it is not.

Phase 1 occurred on the Summer Solstice 1993, with the publication of the Calanais Edition of *Open Creation and its Enemies* by Asger Jorn. Fifty copies of this text were taken to Calanais, located in the Outer Hebrides. By using the Gaelic names of *Calanais*, and *Alba* for that part of Scotland (it is the position of the London Psychogeographical Association (LPA) that the name *Alba* should not be applied to all of Scotland), we hoped to encourage some future art historian to muddle up this Gaelic name with the town in Italy where the *Lettrist*

Internationale attended a Congress shortly before the founding of the *Situationiste Internationale.*

The text was published without a card cover. However the publication also included an envelope equipped with two labels which included a mock first day of issue stamp, and the legend that the envelope was an intrinsic part of the publication. A piece of wool was also put in the envelope, and a piece of this weird plant (whose name I've forgotten and which looked like a strand of wool) was placed where the envelope was sealed. This was a way of detecting the real wankers, who would preserve the envelope intact, hoping thereby to maintain a 'perfect' commodity. As the envelope would bear their name, what few pounds they gained by this strategy of preserving this piece of Book Art intact would be more than compensated by the fact they would be shown to be a complete arsehole.

With Phase 1 complete (which was not a moment too soon, as the Isaac Newton Institute used the same Summer Solstice to stage a three-day orgy of mathematics culminating in their claim of having proved *Fermat's Last Theorem*) we could proceed to phase 2: The English Edition. We planned that many people would assume that the appellation *English* referred to the language in which the text had been published, as opposed to the place where it was published. They would therefore not be aware of the Calanais Edition, which has a completely different preface. This was done following a precept of Sun Tzu's *The Art of War* which states:

> "All warfare is based on deception. Therefore, when capable of attacking, feign incapacity; when active moving troops, feign inactivity. When near the enemy, make it seem that you are far away; when far away, make it seem that you are near. Hold out baits to lure the enemy. Strike the enemy when he is in disorder. Prepare against the enemy when he is secure at all points. Avoid the enemy for the time being when he is stronger. If your opponent is of choleric temper, try to irritate him. If he is arrogant, try to encourage his egotism. If the enemy troops are well prepared after reorganisation, try to wear them down. If they are united, try to sow dissension among them. Attack the enemy where he is unprepared, and appear where you are not expected."
> (*Art of War* Chapter 1)

(When this text is compared with Guy Debord's sad book *Le "jeu de la guerre"* — co-written with Alice Becker-Ho — we have to ask ourselves whether Debord was "feigning incapacity" when he claimed his game was the "solstice of war" which "exactly reproduced the totality of factors which affect war, and more generally the dialectic of all conflicts". Being amongst the few who have played this game, I would advise those thinking of taking the trouble to construct a board and assemble the necessary beads to act as playing pieces, that the game is crap.)

Publishing is war carried out by other means, or if you prefer war is publishing carried out by other means. No matter how you construct the syllogism, both are concerned with imposing will upon the other, whether cold steel or hot lead. Both the general and the publisher "must create a situation which will contribute to their accomplishment" (*Art of War,* Chapter 1) The first four fundamental factors may vary, but the fifth remains the same, *Doctrine.* Having breached the defences of the bourgeoisie with the Calanais edition it was necessary to exploit the momentary (or even transient) advantage with a swift follow through.

The *English* edition was published with two different covers, each bearing the following text:

> "You may discover that this same text is available with a different cover but sporting the same ISBN No. in seeming contravention of the rules as regards the allocation of the International Standards Book Numbers. Be not alarmed! This is not an oversight. The cover does ***NOT*** constitute part of the publication and is merely provided to ensure that the text is not damaged in transit. On receipt of the text, we request all librarians to remove this cover (and indeed to expunge all trace of it, thus ensuring no problems for bibliographers of the future. ISBN No. 1 871 593 20 4"

Did any of the Librarians comply with this simple request? Did they fuck! — "In general, when invading hostile territory, the deeper one penetrates, the more cohesion it brings; penetrating only a short way causes dispersion." (*The Art of War* Chapter 11) Following this non-compliance it has been decided to strike a deep wounding blow at the very concept of the Book, to undermine its conceptual integrity and to create a situation which will advance our doctrine *within the library.* This involves shelving the idea that a book is constituted by a collection of pages somehow assembled together. Such an object is merely a physical manifestation of what is essentially a social relation.

The letters scattered on a page have no intrinsic meaning, but gain their force by relations of convention which frequently stimulate the formation within the mind of the reader, of a set of thoughts analogous both with those of the writer, and even with other readers. By applying the theories of Isidore Isou to the book we can produce the *Lettrist Book*:

> "Each art goes through a period of growth (the amplic phase) and then decline (the chiselling phase). The former is characterised by "pretexts exterior to the art itself: anecdotes (battles, epics, divine struggles), sentiments (romantic) or ideas (philosophical, social etc.)"
> (*What is Lettrism?* by Maurice Lemaître,
> *Ur* No.3 second series 1952)

The latter phase "is the period in which the art turns upon itself and changes its vision from external goals to goals of internal arrangement, proper to art itself (. . .) It is the period of concentration of means and elements to the detriment of the 'story'"(ibid.)

Here the book exists as references to a text which has never been manifested. Both text as a series of characters, and collated pages as a record of that series, become erased from the process, as they are anyway, entirely incidental. The book is now centred around an ISBN number, the registration of which triggers the mighty capitalist machinery into operation. (As Negri says "Capital does not present itself as the organiser of labour power, but as the registration and management of the autonomous organisation of labour power." (*Dalla Transizione al Potere Costituente* in *Derive Approdi* #2, Summer 1993. See *Transgressions #1* for Roberto Bui's important response *Gemeinwesen and Possibility*).

Following registration, the copyright libraries request copies of the book. However, with the new *Unpopular Books* title, *Is the Queen an Alien*? all that they received was a copy of the quote for printing plus some sort of shoe box. An accompanying letter informed them that every effort will be made to ensure that the book gets plenty of reviews, and that copies of these reviews will be sent to them, which they can store in the box until there are sufficient to account for the specified number of pages. (This text will also be included.)

Of course, we don't for one moment think that the librarians will carry out our requests (well, they might). However, hopefully this will cause some sort of bureaucratic problems as they try to sort out how to deal with this latest development in Book production.

However, we are desperately aware that tinkering in this way with the very fabric of the universe can have disastrous results. We have thus devised a way to balance this launching into abstraction with a concretisation of a lost text, to wit Thomas Nashe and Ben Johnson's play *The Isle of Dogs.* This text has been lost for many years following the suppression of the original performance in 1597. (Johnson was thrown in jail and Nashe went into hiding.) There was a copy in the notorious *Northumberland Manuscript* which Spedding found in 1867. This had once belonged to Francis Bacon, and is used by Baconians to support the notion that Francis Bacon penned the works of Shakespeare. Fortunately, the text of this play is missing.

Our goal is to re-stage this play on the Mudchute, a piece of open ground on the Isle of Dogs. This will be done in 1997 to coincide with the 400th anniversary of the original production. Unconstrained by an original text, we shall be able to use this as part of our cultural programme to demoralise the capitalist class.

Richard Essex.

Unpopular Books
Box 15,
136 Kingsland High Road,
London E8 2NS

Sibylle Fraser

Underground Printing in Europe, 1933–1945: An Historical Observation

Almost simultaneously with the invention of movable type, means were sought to restrict the printing and distribution of materials containing views dissenting from those officially held by the leaders of church or state. As early as 1485 Archbishop Henneberg of Mainz demanded that books be examined for their content and the possibility of their containing doctrines in opposition to those of the Roman Catholic Church.[1] This action was a forerunner of the *Index Librorum Prohibitorum* (1559) initiated under Pope Paul IV (1555–1559).

When Henry VIII broke from papal authority and vested church leadership in himself, the immediate flight of many Englishmen faithful to the Roman Church set the scene for the beginning of extensive underground reaction. Ultimately underground printing, as well as "above ground" printing in the countries of refuge, produced a substantial stock of titles for smuggling back into England.

The Anglican Church became firmly established under the reign of Elizabeth I, and dissenters not willing to swear allegiance to the church were again restricted in worship and publishing. The story of dissenters who fled the reign of Elizabeth I and engaged in printing Bibles, prayerbooks, commentaries, sermons and pamphlets for clandestine distribution at home in England, has been dealt with in considerable detail.[2]

This English circumstance, which spawned religious underground printing, has had typical political counterparts from the beginning of printing to the present in such cities as distant as Teheran and Witwatersrand, Santiago and Warsaw—cities where printer and publisher have employed misleading information on the title page or in the colophon in order to diminish the possibility of detection.

Fig. 1. Photograph by Robert Doisneau. From Imprimeries Clandestines *(March, 1945), a literary series published by* Le Point *(Toulouse).*

From recusant times, and even before, to the close of the nineteenth century, clandestine printing was often clumsy in appearance with error and equipment marks being the sign of the hastily produced clandestine book, pamphlet or broadside. A turning point in physical appearance and style could be marked from the innovations of those revolutionaries in Czarist Russia who, at the close of the nineteenth century and down to the time of the 1905 Revolution, introduced color, typographic variety, illustrations and even photographs in their clandestine satirical periodicals and newspapers.[3]

A feeling for the dugout "press room" (fig. 1) reached only by tunnel, lit by fume-producing light, with foul air heavy with risk of detection can be read in Kedrov's account.[4] Although swallowing the protest broadside or leaflet to escape detection is part of the legend of such activity, concern for the risk of the underground newspaper reader may have been pioneered by Lenin with his *Iskra* (1900–1905). Many issues of *Iskra* were printed on exceedingly thin paper which softened when wet allowing reduction in size to that of a small wad.

Although *Iskra* and the satirical periodicals of the 1905 revolutionary period were technical as well as historical achievements, for innovation and imagination, no publishing of a clandestine nature in the first four centuries of printing has rivaled that which was to come in the early decades of the twentieth century. The threat of fascism in Germany and its ultimate reality in the 1930s forced dissenters in Germany, and later in the occupied countries, into the production of a variety of publications; publications which in format, content and design were pioneering.

This brief account which follows is not put forward as a history but rather as a reminder, a half century later, of what extremity did produce in technical and graphic achievement in one recent period, 1933 to 1945. Emphasis is given to camouflaged publications produced by individuals belonging to, or associated with, the German Communist left.

The story of the systematic eradication of Germans holding opposing views to the National Socialists and caught within the borders of Germany in the early years after 1933 is too well documented to need comment here except in the sketchiest of form.

Within the month of Hitler's takeover of Germany (January 1933), the staged burning of the Reichstag was used as the excuse to restrict free speech, free printing and publishing, the right to assemble, and the right to privacy of personal mail. At the same time, a wave of arrests caught those leaders of parties opposing the NSDAP, e.g. the Social Democratic Party, the Communist Party and various splinter parties. To dramatize public policy toward printing material of a dissenting or negatively critical nature, the night of May 10, 1933 was chosen for the now famous book burning in university squares throughout Germany. The works of Freud, Brecht, Toller, Mühsam, Heinrich and Thomas Mann, Tucholsky, Bredel, Lasker-Schüler and numerous others were burned.

Fig. 2. Illegal newspapers being transported in a baby buggy. Somewhere in The Netherlands in the 1940s. Photographer unknown.

By July 1933, a law was put into effect closing the publishing houses and confiscating the printing equipment of labor unions and various political parties. In the case of the Communist Party, the leadership was not caught unaware since this was not the first time their printing efforts had been restricted. The CP leadership saw the "handwriting on the wall" and prepared for going underground as early as 1932. Eugen Schönhaar was chosen to organize the technical framework and recruit designers, typesetters, and printers as well as individuals in the printing support services. In Berlin alone, ten printing houses[5] stood ready to devote their efforts to the "illegal" publishing efforts, while keeping a neutral, printing-as-usual appearance. According to Heinz Gittig, author of the most definitive account of this activity, there were hun-

Fig. 3. Camouflaged cover of an important anthology containing work by Brecht, Becher, Feuchtwanger, Graf, Heym, Heartfield, Kisch, Kläber, Mehring, Toller and others (1935).

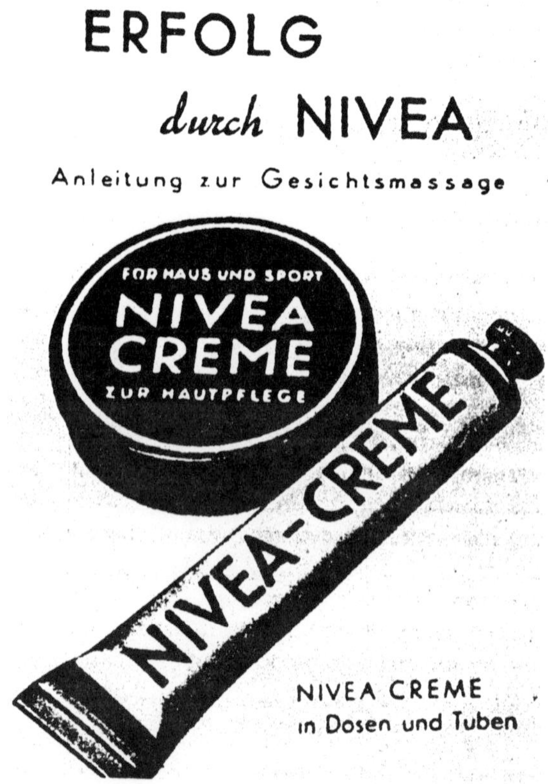

Fig. 4. Camouflaged speech of Maurice Thorez delivered at the Seventh World Congress of the Comintern, 4 August 1935. Also contained are: "Several Hints for Communists and Anti-Fascists."

dreds of printers, throughout Germany, willing to "do their share", if needed. Schönhaar and three other organizers of the printing effort were caught and shot on February 1, 1934.[6]

With this turn of events, the CP and the Social Democratic Party began setting up a network of underground printing operations, although the leadership of the CP left Germany in 1933 and directed the program first from Paris, then Prague and finally from Moscow. Funds were supplied from the Soviet Communist Party for the maintenance of the German project. A skeleton leadership remained within the borders of Germany to supervise organization and distribution.

It should be stated, at this point, that underground printing in Germany was not confined solely to the CP. Social Democrats, as mentioned above, and smaller parties were attempting to keep a free spirit alive. But in resources, human and financial, the other parties could not compete in quality and in more innovative areas such as the camouflaged publications for which the CP underground became known. Nearly eighty percent of all such camouflaged, anti-Nazi material, produced within Germany and in exile during the twelve years of the Third Reich, was done by the CP.[7]

Discovery by the police and Gestapo was inevitable for many within the opposition ranks. One of the primary ways of avoiding detection lay in the system by which each person working for the CP, whether in printing, editing, publishing or distribution (fig. 2), had a limited field of responsibility and little contact with anyone else. In this way, if manuscript, galley or a cache of publications were found in someone's possession, the individual being questioned would be unable to provide information leading to the arrest of others. Thus, only through patient sur-

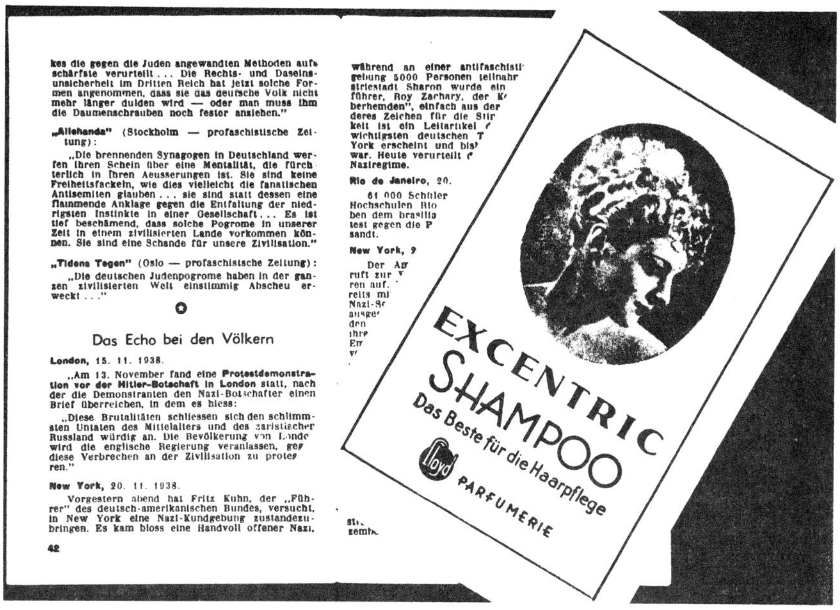

Fig. 5. Shampoo package with plain wrappered pamphlet (1939) opened to pages showing world press reaction to the plight of the Jews in Germany following Kristallnacht.

veillance and close examination of the printed matter being distributed, was detection possible. It is surprising then to see that the first few efforts of the German Communist Party in underground printing were without attempt to disguise the content. The publications were simply printed in straightforward form or with plain wrappers. This changed quickly. The simplicity and obvious advantages of camouflaged covers in reducing risk were adopted already in 1933. But the choosing of just any cover was not sufficient. Success was to be found in reproducing exactly covers of small pamphlets, manuals, and mass produced literary texts already well known and widely distributed (fig. 3). There was little likelihood that, initially, authorities would be suspicious of manuals on sewing machine repair, booklets on the care and feeding of canaries or cacti, pamphlets on steps to more beautiful hair or skin (fig. 4). Shampoo was, at that time, sold dry in small envelopes with an illustration or design on the front and directions on the back. The reprinting of these packets, with a plain wrappered pamphlet printed on thin paper included inside, was a particularly successful way to camouflage a message (fig. 5).

Another noteworthy design disguise was the pocket book, such as the Reclam paperback series of literary classics published in Leipzig for more than a century. Reclam books were a staple of the student population for years. Vast quantities of them were in circulation in the new book and antiquarian trade. The camouflaged Reclam publications were first designed to look, more or less, like the originals but, through sad experience, even small deviations in color, typeface or paper led to detection. Subsequent camouflaged publications appeared in covers indistinguishable from the originals.[8]

Many publications were produced in plain wrappers. Of the twenty-seven titles known to have appeared in 1933, five were plain.[9]

In general, until 1935, underground printing was carried on both in and outside of Germany. But from 1935 on, most printing was done outside of the country because of the danger. Also, by that time, most of the presses, within German borders, had been discovered or the printers had been imprisoned. Thus, the central headquarters of the German Communist Party in Moscow decided to build networks among Germany's neighboring countries and direct the distribution from there.

Each part of Germany was supplied with underground printing according to a specific and tight distribution network. Berlin-Brandenburg and central Germany (Dresden, Halle, Leipzig) received the illegal shipments of printing mainly from Czechoslovakia. The north of Germany (Bremen, Hamburg, Hannover and Kiel) was supplied by shipments from Denmark and Sweden; the northwest, from Belgium and The Netherlands; the southwest from France and Luxembourg; the east from Denmark, Sweden and Czechoslovakia.[10]

Ingenious ways were developed to transport these materials into the Reich. Some shipments came by private boats, some by commercial cruise ships from Scandinavia. In some cases, shipments were floated down rivers in watertight bundles, at night, to be picked up at prearranged times in a certain locality. The bulk of the materials was intended for the large industrial centers, always a stronghold of the CP and others within the sympathetic left.

The primary purpose of these pamphlets was to keep members of the German Communist Party informed of the work of the Party outside the borders of Germany. This was done by publishing the proceedings of Party congresses, accounts of political events and party interpretations of these events, e.g. the Spanish Civil War and the role of the Soviet Union and other nations in that conflict. There was also a strong indoctrination message present in many of the writings.

The Roman Catholic Church was one organization the CP tried to draw into cooperation since there were factions in the Church active and working in opposition to the Nazi regime. Although the Church rejected an alliance with the CP, some pamphlets contained articles which sought a closer cooperation.

The subject matter addressed in the publications was as diverse as the camouflaged format. The argument that Hitler's policies would harm the worker and the farmer and that fascism should be fought under the leadership of the Soviet Union received much attention. The latter proposition, however, lost support after the Hitler-Stalin Pact of 1939.

Other subjects treated were the problems of the labor unions in the Reich, the issue of discrimination and subjugation of women under Nazi policy, accounts of political prisoners in concentration camps. Some pamphlets were directed to the youth of the Reich giving details of youth congresses in different parts of the world. Information on the work of communist parties in other countries, e.g. China, Hungary, etc., was regularly given. Even parodies of Nazi songs were included from time to time. Curiously, there was little mention of the racial issue. A one-page statement to the effect that "Jews are people like we are" appeared in a publication in 1935.[11] In that same year, two three-paged items appeared titled "The Racial Question,"[12] and "The Jewish Child."[13] Speeches of Stalin, G. Dimitroff, a leader of the Comintern, Walter Ulbricht, Wilhelm Pieck, leaders in the GCP, were also published.

In short, these small (average size 12 cm. x 8 cm.) booklets of twelve to thirty-six pages remain more than a bibliographical curiosity. They are primary documents revealing a sophisticated news and indoctrination medium, edited by professionals, and conveying information on world events in light of the Comintern and GCP policies. Creative literary work was infrequent although a few essays, commentaries and plays, in opposition to the regime, were included by such well-known writers as Gustav Regler, Thomas and Heinrich Mann and Lion Feuchtwanger.

Literary and graphic creativity appeared more often in the less ideologically conforming underground printing of other countries, for example The Netherlands.

The quantity of these German communist camouflaged publications is of some interest. The table below indicates something of the impact of the war on their production:

Year	*Number of Titles*[14]
1933	27
1934	102
1935	70
1936	75
1937	67
1938	60
1939	54
1940	2
1941	1
1942	8
1943	9
1944	9
1945	0

Following the initial year, pre-war production was relatively stable. Once World War II began and the distribution network disrupted, production nearly came to a standstill. As the war progressed, domestic circumstances became more disorganized. The opportunity to smuggle material of a printed nature across borders increased and the printing activity revived somewhat.

The content of the camouflaged pamphlets did, however, change considerably in the wartime and not surprisingly. Advice on how and when to listen to illegal radio stations, warnings about the disintegration of the country, and calls to resistance were noted in the three publications for 1940 and 1941 held by the Deutsche Staatsbibliothek.

By 1942 some pamphlets took an open position on how to end the war and in 1943 two curious pamphlets appeared with signs of the zodiac and articles on "What the German woman wants to find out from the stars," "Occurrances on the planet earth," "Dangers Ahead," "Götterdämmerung," the "Battles of 1939–1943," and "Statistics of the war." This was the year 1943. The successes of the war turned against the Germans. This was the year of the Battle of Stalingrad. In this same year, articles also appeared on how to simulate different ailments to avoid being drafted, how to acquire pharmaceutical products and how to make rubber stamps in order to falsify documents.

In 1944, articles emphasized the destruction of the Reich through bombings as well as a continuation of articles on simulated illnesses.

In the closing months of the war, when consumer goods were in short supply, the depiction of consumer goods on the covers as a means of camouflage was discontinued for the simple reason that the publications would then be conspicuous. Instead, pictures of Göring, Hitler and other propaganda covers were used.

Unfortunately, because of the secret nature of this printing wherever it took place, specific details of people, printing plants, equipment, means of supply, are not recorded in print. However, information is available from an ever diminishing number of participants, some of whom are reliable informants. One may also deduce from the quality of the printing in the pamphlets, the variety of display type, the evi-

FEVRIER 1941 · No.2

VALMY

NOTRE DEVISE : Un seul ennemi l'envahisseur

LE CANON DE GENES

Le neuf février, à l'aube, une puissante escadre anglaise bombardait le port de Gênes elle y détruisait des installations et des entrepôts, touchait des établissements industriels, désorganisait les préparatifs d'une invasion possible de l'Afrique du Nord.

Cette brillante action navale démontre plusieurs vérités de fait.

D'abord, la flotte britannique reste maitresse des eaux non seulement dans l'Atlantique où son pavillon flotte à peu près seul si on excepte [illegible] alliés, mais encore dans [illegible] [illegible]

Ses côtes allongées, ouvertes à tous les risques d'invasion, n'auraient pu se défendre des coups qu'on y eut [illegible]

[illegible] tout cela qui fait entendre le canon de Gênes

Mais ne désespérons point. Les puissances de l'Axe, aujourd'hui sur la défensive, peuvent encore porter des coups violents. Elles ne peuvent posséder la mer qu'avec notre complicité ! Cette idée d'une trahison succédant à la capitulation est odieuse à tous les Français.

Au contraire, nous attendons le jour où de vrais chefs reprendront la lutte pour reconquérir la terre de France, souffrante et pillée, que nul ne peut abandonner sans forfaiture.

VALMY

«La Révolution périt non pas pour avoir trop voulu, mais pour n'en pas avoir assez osé !

LAMARTINE

(Histoire des Girondins)

Fig. 6. This resistance paper was printed, in its early issues, with a child's rubber printing set. The lock-up device held four lines of type.

dence of cover illustrations produced by photo offset and the variety of type sizes available—that relatively established printing plants were used for the production of these clandestine documents. We may conclude from the well-bound and cleanly trimmed work that the majority of the camouflaged pamphlets examined were produced by skilled workers. This printing was not, for the most part, done in romantic circumstances such as the physically underground sites alluded to earlier. Certainly some publications were produced in difficult circumstances including those in occupied areas where, in some instances, the controls were comparatively relaxed.

*****

A comment on the publications and printing operations in occupied areas provides additional insight into printing under extreme conditions during this twelve-year period of upheaval.

The nature of underground printing activity in those countries under German occupation tended to follow a similar pattern of collabora-

Fig. 7. Det Frie Norge *(Free Norway) camouflaged as a chess player's guidebook. The actual cover is pictured at right.*

tive effort involving risk at every level from the "requisition" of increasingly scarce supplies, the actual printing activity, movement of presses to less visible locations and the distribution process. Often plants engaged in printing permitted materials such as occupation government newspapers, books, and job printing while carrying out the printing of dissenting literature in nearly "full-view" of the occupying officials. Denmark, France, and for a time, Czechoslovakia are countries where this "above ground" printing was known to have taken place. In other countries, such as Norway and The Netherlands, it was safer to print in physically disguised facilities and on manually operated equipment. Doubtless, there are exceptions to both generalizations.[15]

Belgium had an active underground newspaper effort. It is claimed that approximately two hundred titles were published, some only a few pages and in small editions. However, the major underground paper, *La Voix Des Belges*, reached an edition size of 50,000 forthnightly.[16] Churchill's admiration of this effort led to the naming of yet another paper as "Churchill's Gazette." Many of these papers, published by non-partisan groups, confined their content to news of the war and articles designed to encourage the general population.

The situation in France was more complex because of the division of the country into occupied (North) and unoccupied zones (South). Many involved in the printing trades in the occupied part became a compliant tool of the occupation forces. The well-known division within the population as to who the real enemy was and against whom the fight should be directed created a further impediment to an effective underground movement in the early period of the occupation. The Communist press found itself in the familiar dilemma of many a national communist party following the Hitler-Stalin Pact and directed its attacks in its underground efforts mainly against the Vichy regime. In general, during the first years of the resistance, underground printing coalitions expended considerable effort in establishing a press apparatus free from the taint of the Vichy regime and their publications.

French resistance publications were mainly

Fig. 8. A die-cut Dutch underground pamphlet which announced on the envelope in which it was distributed: "A look in the new order." When closed, as shown above, Hitler is bringing food to a starving mother and daughter. The picture is captioned: "This is the new order" (mid-1940s). As the pamphlet is opened (below), Hitler is, in fact, removing *the food from the table and taking it to one of his men.*

newspapers in relatively small format with an edition size ranging from as small as fifty copies in the case of the *Valmy Gazette* (fig. 6) to as many as 300,000 in the case of some issues of *Défense de La France*. The content of these papers tended to be news of the war effort as well as world news taken from the BBC broadcasts. As the victorious forces came closer to France, the number of papers proliferated, and the more common it became to see offers in print to work with the victors.[17] By the close of the war more than 1,200 newspapers had appeared[18] and, according to Marie Granet in "A Short Manual of Underground Printing,"[19] approximately

1,500,000 copies of these various newspapers were appearing per month in 1944.

Norway had an active underground publishing effort and effective efforts were carried out from the beginning of the occupation in the spring of 1940 until 1945. Opposition publishing ranged from openly published works containing camouflaged satire (one of the most famous of which was the story of Snorre the Seal)[20] to camouflaged periodicals such as *Det Frie Norge* (Free Norway) (fig. 7) published in London and distributed throughout Norway. The documentation of Norwegian underground printing and publishing began early with such extensive compilations as Jensson's multi-volume *Norsk Presse under Hakerkorset; den "Illegale" Presse*[21] and Kildal's *Presse- og Literaturfronten under Okkupasjonen 1940–1945.*[22] Much of the Norwegian underground printing was done with hektograph equipment and various types of office reproducing machines which could be sequestered and serviced more easily than conventional presses.

The Netherlands was the only occupied country that managed to maintain a wide range of conventional printing and publishing activity including "fine press" books on handmade paper with original graphics. The story of Dutch underground printing has been told and re-told by de Jong,[23] Simoni[24] and others and examples appear, from time to time, in exhibitions of printing, particularly the broadside poems which were a feature of this country's resistance printers. These broadside poems, short stories, and plays were the literary substance of the Dutch phenomenon with proclamations, religious writings and portions of the Bible making up the rest.[25]

The Dutch printing resistance effort (fig. 8) was often subtle and, in general, lacked a certain ideological edge. Poems were often in commemoration of an historical event in which the nation took a stand against an oppressor of the past. Other writings are in memory of patriots of other times. The underground *belles lettres*, as well as the Dutch periodical press, often lacked the strident and aggressive tone of the French and Belgian press. Its wider interest to us here is in the unusual quality of the typography, the apparent access to printing equipment and good quality type. The quality and variety of paper on which Dutch work of the period was printed is unmatched in the work produced in other occupied countries.

In conclusion, printers, service personnel, publishers, distributors and readers in the occupied countries operated at risk just as did the German resistance. Their lives were lost as often as their German counterparts. All participants, regardless of location, remain in the direct line of those oppressed and martyred printers and publishers of earlier causes and are worthy of our recognition in this period of worldwide threat to the democratic press a half-century distant from those times.

Notes

Over the past twenty-five years I have examined camouflaged and other clandestine publications in my own collection and in institutional collections. It has been possible for me to interview in this same period a number of individuals with first-hand, as well as reliable second-hand, information. Among these individuals have been Heinz Gittig, Michael Katanka, Age Abma and Walter Goldwater. Several individuals have preferred to remain anonymous.

1. S. H. Steinberg, *Die schwarze Kunst.* 2nd ed. (München: Prestel Verlag, 1961), p. 311.

2. Leona Rostenberg, *The Minority Press and the English Crown; a Study in Repression, 1558–1625.* (Nieuwkoop: B. De Graaf, 1971), 263 pp. A collector's account is given by Charles A. Fracchia, "Racks, Rascals and Recusants, the Search for the Hunted Book," *Book Club of California Quarterly Newsletter* 32 (Fall, 1967): 75–83.

3. An indication of the variety of these satirical journals can be seen in E. P. Gomberg's *Russkoe isskustvo i revolvutsia 1905. Grafika i zhivopis.* (Leningrad: Leningrad University, 1960), 172 pp. and in: E. P. Demchenko's *Politicheskaia grafika Kieva . . . 1905–1907.* (Kiev: Naykova Dymka, 1976), 187 pp. Color reproductions of these journals, with limited English commentary, can be seen in David King and Cathy Porter, *Images of Revolution.* (New York: Pantheon, 1983), 128 pp.

4. M. Kedrov, *Geheimdruckereien.* (Hamburg/Berlin: Carl Hoym, Nachfolger, 1931).

5. In Heinz Gittig's *Illegale antifaschistische Tarnschriften 1933 bis 1945* (Leipzig: VEB Bibliographisches Institut, 1972) on pages 46 and 47, he lists the ten printing houses which had cooperated, were shut down by the Gestapo and were listed in the Gestapo's report on 14 March 1934. They are: Buchdruckerei Otto Müller, Grosse Frankfurter Str. 15; Druckerei Karl Röhrig, Holmannstr. 18; Buchdruckerei Franzke & Grunack, Prinzenallee 83; Buchdruckerei Neumann und Mattert, Potsdamer Str. 134a; Buchdruckerei Schuster und Pätzold, Eisenbahnstr. 13; Buchdruckerei Grossmann, Neanderstr. 4; Buchdruckerei Sparsam und Wolf, Stallschreiberstr. 27; Buchdruckerei Grossemigen, Eichwalde, Druckerei Dzubas, Blücherstr. 31; Setzmaschinenbetrieb Albert Volz, Lindenstr. 29.

6. Brief mention of Schönhaar is given in a number of sources including Institut für Marxismus-Leninismus. *Geschichte der*

deutschen Arbeiterbewegung. Vol. 5. (Berlin: Dietz Verlag, 1966), p. 30. But this source, like the rest, gives virtually no background on his life.

7. Gittig, p. 15.

8. In addition to Reclam as an imprint which, according to Gittig, was used as a false publisher in thirty-two instances, Verlag für Kunst und Wissenschaft, Albert Otto Paul (Leipzig); Hachmeister & Thal (Leipzig); Bibliographisches Institut (Leipzig); Ensslin & Laiblin (Reutlingen) and Hyperion Verlag (München/Berlin) accounted for the greater percentage of use.

9. Gittig, pp. 111–115.

10. Interview with Gittig and others.

11. In a publication camouflaged as Walter Pötsch, *Die Grundlagen des jüdischen Volkes. Eine notwendige Abrechnung*. (Breslau: Pötsch-Verlag, 1935). p. 3.

12. In a publication appearing in plain wrappers. (S.l., 1935), pp. 20–23.

13. In a publication camouflaged as *Deutsch für Deutsche*. (Leipzig: Verlag für Kunst und Wissenschaft, Albert Otto Paul, 1935), pp. 111–114.

14. Compiled from statistical data presented throughout the bibliographical portion of Gittig, pp. 111–237.

15. Interviews with individuals mentioned in initial note.

16. Belgium Ministry of Information, *The Underground Press in Belgium*. (London: Lincolns-Prager, 1944), pp. 6, 11.

17. Reinhard Freiberg, *Die Presse der französischen Resistance, 1940–1944*. Dissertation. Berlin: FU, 1962.

18. See items 1–1015 in *Catalogue des Périodiques Clandestines Diffusés en France de 1939 à 1945*. (Paris: Bibliotheque Nationale, 1954) and the supplement which contains additional titles numbered 1087 to 1106. Items 1016 to 1086 are foreign entries.

19. *Lithopinion* 3 (1968): 11.

20. Kari Skjønsberg, "Snorre the Seal: A Children's Book in the Norwegian Resistance Movement." *Phaedrus: An International Annual of Children's Literature Research* 8 (1981): 18–19.

21. Gunnleik Jensson, ed. Oslo: Tell Förlag, 1945–46. 3 vols. various pages.

22. Arne Kildal, *Presse- og Litteraturfronten under Okkupasjonen 1940–1945*. (Oslo: H. Aschehoug, 1945), 256 pp.

23. Dirk de Jong, *Het vrije boek in onvrije tijd*. (Leiden: Sijthoff, 1958), 341 pp.

24. Anna E. C. Simoni, *Publish and be Free*. (The Hague: Martinus Nijhoff, 1975).

25. *Ibid.*

The beginning of Mark from the Book of Durrow

Who else....

Oh yes a girl I used to know.

At last- The Final Party.

SUN MOON AND STARS PRESS.

BY ANDI McGARRY.

&

&&

&&&

&&&&

&&&&&

&&&&&&

&&&&&&&

&&&&&&&&

&&&&&&&&&

&&&&&&&&&& &c

vertigo publications

&c

bob cobbing

"THE GOOD GRAY POET": WALT WHITMAN AND THE BOOK

Sarah Jacobs

Poets devour books. You may like to think I mean this metaphorically. Poets deflower books — I am thinking of Mallarme's rapacious eye. Poets deface books. They scribble in margins, rip out pages, play havoc with their neighbours' forms. How can I tell you about Walt Whitman? All above categories apply.

I am sure you remember the fellow. His bearded image stares out from every edition he published, printed or procured — and there were nine editions of *Leaves of Grass* between 1855 and 1892, mostly self-published, to say nothing of editions of his other poems and the prose works. Walt Whitman, his signature, 'scrawled with big coarse ragged pen'. Starting jaunty, he ages quickly (having gone white) and can't make up his mind whether he prefers 'the laughing philosopher' or 'poor, old, blind, despised and dying king.' *Self with butterfly* remains a favourite. For the title portrait of the 1888 edition, an austere profile , he asks for the curls to be combed out — obviously an early plea for an airbrush — but for some hair to be left though 'not in a top knot or Italian curls, which are not and never were .... '.

The text was intended to express his Self — he was the Bard of Personality — and the images were consciously placed to counterpoint the text. If we are to claim him as avator, it is as well to recollect he may have been a pioneer in calling an edition 'new' merely because it had a fresh frontispiece or photograph coupled with a different binding. He was certainly credited, by his contemporaries at least, as being a pioneer of self-publicity, composing his own puffs (the 'writings are a powerful sense of physical perfection, size, health, strength and beauty, with great amativeness, adhesiveness, a wonderfully buoyant joyousness .... ') while Emerson's unsolicited praise for an early edition is actually lettered on the spine of the successor volume — without the other poet's consent. He placed his own articles in acquiescent publications:- 'The newspapers still keep up their talk about Walt Whitman .... ' Some of what was printed seems to have been disinformation, for example that his volume of prose had sold out before publication.

The books he published, he sold and shipped himself, keeping details in careful ledgers. 'Though personally in solitude I am travelling all over the world.'

As well as the bound books (the multiples), he made notebooks and scrapbooks, and he routinely annotated books: *The Art of Abbreviating Those Studies Which Give the Greatest Labor to the Memory.* He used the very best paper for the little pocketbooks stitched by friends or sewn with pink tape on which he wrote hospital notes during the Civil War, while the 1860 *Leaves* was composed mostly on pink sheets cut from wrappers left over from the 1855 edition, and on blue tax forms, though when he moved to the Calamus series — the manly love poems — he eschewed pink for white wove. Fragments were written on wall paper or the insides of envelopes, then clumped into bundles, or left strewn about the floor. A contemporary refers to litter. You can tell he has in mind a pig stye.

He composed like a constructionist, moving his lines around like so many planks. Another contemporary records seeing him twisting a loop of string from which hung scraps of paper, like dangling keys. Each was the idea for a poem. So in his poetic process physical manipulation went hand in hand with writing. He composed like a collagist, pasting and layering the paper. He wrote in blue, brown and black ink and, of course, red. Crayons and pencils in many colours were also used. Any one page might contain one or any combination of these. "He preferred to write with pens and pencils which were, literally, enormous, and he always wore a broad-brimmed hat." This seeming non-sequitur from a devotee seems apt if one considers the contents of a scrapbook:- He cut book pages down to stubs, and pasted in green wrappers (another left over from the first edition of *Leaves of Grass*) cut to size, unused tax forms, endpaper from the second *Leaves*, and clippings about fairy tales, Indian names, learning to read by phonotypy. On yellow scrap pasted to white scrap 'Mikkel Angelo'; 'Apollo Bel-vi-dar.' On three white scraps pasted to a green sheet:- '*Myrathas,* in old persian mysteries was the name of the sun; *Mylitta,* that of the moon (the earth); (rondure) 'this huge *rondure* on which we rest.' Gilson — Beans. Cor. Sudbry and Court adv. of a hog hitched to a carriage.'

Invariably he talked about his work in terms of making, shaping, moulding or even licking into shape:- 'A living, flush, eating and drinking man, — the mould — and as from that, without wincing, to mould a book.'

However, we should not be under an

I will make the poem of materials

The cumulous character of the book is a great factor, perhaps even the jaggedness, or what might be call'd so from the conventional and tidy principles of 'art'.
probably is so anyway, bad days and nights, no hour without its suffering

My book and I, what a period we have presumed to span

illusion that the handwritten books represented anything like an attempt to create a unique artist's book. His appreciation was all for the final product, praising not only printers but proof readers:- a class to whom 'all the hundred of writers and all the .. readers are unspeakably indebted.' They extricate the text from 'the mass of chaos ... excesses or balks or hiatus and frequent messes of illegibleness and tautology.' He did not hesitate to work over his own hand-books, obliterating, and starting again upside down and over. He seems to move close to some idea of an artist's book in relation to the printed books — wistfully longing for the day when the author ' will do it ... He will some day make his own books, cover and all.' This casts a gloss on his famous explanation — that he was self-published because no established publisher would touch the work, and, in any event, three times running rascally agents embezzelled the takings. He obviously liked making books.

The size of the volumes mattered to him. He teased his brain 'with visions of a handsome little book ... a dear, strong, aromatic volume, ... for the pocket. That would tend to induce people to take me along with them and read me in the open air. ... To have such a book produced in every way according to a feller's simple and unimpeded humours — that has been my idea, is still my idea.'

'I lead no man to a dinner-table, library, exchange,
But each man and each woman of you I lead upon a knoll,
My left hand hooking you round the waist,
My right hand pointing ....'

But he is said to have particularly favoured a small quarto size.

Binding mattered to him, and if he wanted his covers to be plain, we must see this as part and parcel of his image ('not intended to be polished off — left purposely a little in the rough' rather than 'foxy and professional' or 'amiable ... harmless well-fed sleek' — this last of Edmund Gosse whom he compared to a well ordered parlour, crowed with statuary, bric-a-brac, and — of course — a wealth of books in gilt and morocco). There were at least 12 different bindings for the 1860 *Leaves* alone and the correspondence is rich with mentions of green cloth, half-calf, good strong paper covers. While he was not above using gilt for 'a special edition for holiday purposes' he insisted that the centennial edition be 'markedly plain and simple even to Quakerishness ... no sensationalism or luxury — a well made book for honest wear and use and carrying with you; a little bound volume and I fancy you saying 'It looks plain and common enough — not handsome or presentable evidently at first' — but let us see what obviously moved him and there are references to 'robin-egg blue colour with white', and roan volumes, while 'dark-red colours, complete the final technical blowout.' He wanted a 'soft wine colour — dark, pure — for the little book.' We can easily understand how the same inspired fellow could be at one and the same time the erotical bard and 'the good gray poet.'

Type mattered to him and a notebook contains samples of English runic, pica rustic, scribe:- 'the typographical show of my poems — how they shall show (negatively as well as positively) in the black and white page — is always my idea in making them."

His aim above all was for everything to be clear and readable and he fancied wide margins. He preferred the English style:- 'American print is all gloss, glare: the paper is bad, the ink anemic — pale in the gills: makes me think of the prettiness of drug stores — the polished bottles, the painful glitter.'

Whitman worked to the page. He had an idea how a page should look, so he used the lines from his notebooks like filler, closing up unwanted blanks:- 'Printers send word, too, there is a blank here to be written up — and with what?' He is reported as preferring to 'cut off a precious paragraph, ... [rather] than leave the eye offended. So, too, would he accommodate poem to circumstances.' Faced with a section ending on a full page (something he abhorred) without hesitation he took thick blue pencil and excised a chunk. Yet what was important was the *Leaves* comes down 'intact, unbroken, not a comma removed from first to last.' This concern for the integrity of the body of his book shows in the words he uses to describe changes — whether made by printers' error — or by censorship: mutilation, disfigurement, horrible dismemberment. His vehemence should not astound. 'This is no book, / Who touches this touches a man.'

Anecdotally we learn that when asked to correct and guide a book about him written by someone else (Dr Buck) he attacked the proofs 'like a terrible surgeon with the knife and saw and saw the patient.' It is gratifying to learn that he considered the final product to be:- beautiful, vehement, orbic, adherent to primary spinal laws, and (above all) satisfactory in its type, style and paragraphing, exhibiting typographical cleanliness and propriety.

His love for the page and his hatred of censorship come together when he considered how to foil the Attorney General, who sought to expunge whole pages (Whitman, who was widely reported never to laugh, nonetheless seemed to manage a chuckle at being known as that writer of dirty novels). In a copy of *Leaves of Grass*, now lost, he marked lines and passages 'to

The thin red jellies within you or within me, the bones and the marrow in the bones

'How I made a book, or tried to'

chilled with cold types, cylinders, wet paper between us

come out, and their places to be exactly filled with other matter — so they will superficially present the same appearance as now.' Always practical he comments to a publisher (James R Osgood & Co) that the changes should not cost more than 5 or 10 dollars. Although a publicity hound he exhorts: 'Let this whole matter be kept quiet in the house — no talk or information that may lead to newspaper items — the change to be just silently made — the book, and at casual view all its pages, to *look just the same* ....' [My emphasis.] Whatever interest these passages may have for the literary scholar or libertarian, to the book artist they confirm that for Whitman words equate with stuff.

The censorship row was not his only brush with the official mind. He also had a dispute about the price of posting books, more specifically, manuscripts. He called on Webster's dictionary to confirm that a pamphlet is a small book 'consisting of only one or a few sheets stitched together, and not bound — ' and concludes that all literary manuscripts [by inference, this applies to single pages] are book manuscripts and when printed they become 'Books.'

I do not want to cut about for your cursory delectation the poetry and literary work, and it is clear Whitman would not have wanted me to do so:- 'I shall have my friends know that I don't endorse .... miniaturing from Leaves of Grass. No one can know Leaves of Grass who judges it piecemeal: it makes no revelations to the merely literary eye' — a passage (with emphasis supplied) which I take as both warning off, and confirmation. However what is striking is the intermingling of sense of person, and volume. He hits this head repeatedly:- 'O friend, whoever you are, at last arriving hither, accept from me, as one waiting for you at this entrance, welcome and hospitality.

'This is no book — but I myself, in loving flesh and blood. I feel at every leaf the pressure of your hand, which I return, and thus throughout upon the journey
linked together will we go.'

'What you are holding is in reality no book, nor part of a book,/ It is a man, flushed and full-blooded — it is I —'

'Indeed this is no book, but more a man, within
whose breast the common heart is throbbing so
much. No printed leaves, but human lips, O
friend, for your sake freely speaking.'

An English woman, Anne Gilchrist, took the trope literally. Publishing first at the request of Wm. Rossetti (who prepared a truncated English edition which appeared in 1867) 'A woman's estimate of Walt Whitman', she spent three years dreaming of Whitman, only to feel crushed 'when the Book came but with it no word for me alone.' She wrote him an ardent love letter, seeking to nominate herself as his mate. Today, when Whitman is widely known as a homosexual poet, we put one construction on his inevitable refusal. I would like to suggest another. That he did truly reserve himself for the published volume. He replied after a pause:-' ... feel no disappointment .... My book is my best letter, my response, my truest explanation of all. In it I have put my body and spirit.' His repeated references to his conviction that he was writing a Book, one book, a bible ever; his fiddling with editions and with facsimile signature, with individual lines, seems to evidence an identity between body and text, or body and volume, which might be termed a confusion. In contrast to his official (published) works, the correspondence and notes tend to seem pernickety and flat if considered as mere words, though as I have indicated above, they leap to life when considered with the artist's eye, even when seen in black and white copy:-

'The pages of the lesson having writ to train myself — to you I bring them here and now resign with all their blots, to image back the process for your use'

I don't think he had much of a private life — certainly his story is devoid of incident — and I think the poet of personality was right in understanding that he actually had none — if by personality we mean some notion of flamboyant or exclusionary behaviour. He deplored those poets who found themselves

'... a class set apart, circling only in the circle of themselves, modest and pretty, desperately scratching for rhymes, pallid with white paper ... [He exhorted to] recognize with joy the sturdy living forms of the men and the women ... , the divinity of sex, the perfect eligibility of the female with the male .. the eye for forms, the perfect models of made things ... [to] recognize with demur those splendid resistless black poems, the steam ships ....'

Curiously, having taken the decision to go public about 'love flesh swelling ... limitless limpid jets of love' or the smell of his own armpits 'finer than prayer' he seemed to have nothing left for the private — sending, for example, to a Mr. Young:-

A Font of Type
This latent mine .. not non pareil, brevier, bourgeois,
long primer merely,

I guess I am mainly sensitive to the wonderfulness and perhaps spirituality of things in their physical and concrete expressions

The blank left by Words wanted, but (not yet adopted) unsupplied (have) has sometimes an (ghastly, unnameably putrid (a most frightful indescribably)
(It is a bitter) cadaverous meaning. It glares (is) talks arose (?) louder than (sounds) tongues.
(What battl biter taste is throws) Whaat (an acrid a rank unsweetened) stinging taste is (yet) left in (words, all)...

These ocean waves ....

This marked, 'Personal — don't print!' The sexuality of his work is explicit, the homosexuality scarcely veiled, but why hide his love of the paraphernalia of the book?

The strange elision is echoed in a conflict between his love of seeing his name in print and his repeated attempts to remain anonymous. Asked why he had left his name off the title page of an early edition, he said it would have been putting his name on the universe. *Leaves*, he said, was anybody's who chose, as truly as it was his. Anonymity becomes signature. Asked whether he thought 'Walt Whitman' — the name — on a later title page was too big, he shook his head. No. He had had his doubts, but they were only momentary but now he could see clearly that it was about right. They had it now. In that book Walt Whitman was everywhere. Besides, the book was large. It had a fair bulk. It would stand the racket. If he had said 'I, Walt Whitman' in his poems and the text meant only what it said — well then the situation would be sad indeed. But another time he started to quiz himself about the lettering. 'Is it too large? Can it be too large? Does it betray — eccentricity?' and looking at his friend he said he guessed that for this book the name was not too large.

I would hate to leave off without reminding you that Walt Whitman doted on himself — 'there is that lot of me, and all so luscious.' He found the edition of *November Boughs* excellent except for the lettering of the cover:- 'weak pea soup, dish water.' A friend's book came to him 'so nice and fresh like a new pot-cheese in a clean napkin.'

Were you thinking that those were the
words — those
upright lines? those curves, angles, dots?
No, these are not the words — the
substantial words
are in the ground and sea,
They are in the air — they are in you.

Were you thinking that those were the
words —
those delicious sounds out of your friends'
mouths?
No, the real words are more delicious than they.

Whitman was an invalid for many years. Reading today the memoirs of his friend, Horace Traubel, who kept a careful note of all conversations between mid July 1888 and end October ("With Walt Whitman in Camden"), one might almost speculate he had a form of Alzheimer's disease. He was clearly forgetful, kept repeating himself. He spent hours rummaging round his papers — asking Traubel to read out bits. It seems as if often only the thought of his books — the actual physical books — and his preparation of new editions, was sufficient to animate him. Traubel, quoting his words, sums him up:-

"Eyes large with desire, he takes the sheets in both hands...' that is the book — the real, living book!' at the same time turning it over and over fondly and in a spirit of undisguised exhilaration. '... the deed is done! My book, your blood, went to the making of this book! Some men go to the North Pole to do things — some go to wars — some trade and swindle: we just stayed where we were and made a book.' "

Sources

Greenspan. *Walt Whitman and the American Press*, Cambridge University Press, 1990

Traubel, Horace and others, *Re Walt Whitman*, David McKay. Philadelphia, 1893.

Traubel, Horace, *With Walt Whitman in Camden*, D. Appleton and Company, New York, 1908.

Whitman, Walt, *Poems*, Edited and selected by Rossetti, William Michael, John Camden Hotten, London, 1868.

Complete Prose Works, G.P. Putnam's Sons, London, 1898.

Whitman's Manuscripts: *Leaves of Grass* (1860). A Parallel Text edited with Notes and Introduction by Fredson Bowers, University of Chicago Press, Chicago, 1955.

Leaves of Grass. Facsimile Edition of the 1860 Text. Introduction by Roy Pearce, Cornell University Press, Ithaca and London, 1961.

Walt Whitman's Workshop: A Collection of Unpublished Manuscripts. Edited and with Introduction by Furness, Joseph, Russell and Russell, Inc. New York, 1964.

Leaves of Grass. Reader's Edition edited by Harold W Blodgett and Sculley Bradley, New York University Press, 1965.

The Collected Correspondence. 7 vols. edited by Edwin Haviland Miller, New York University Press, 1961 — 1977.

Daybooks. Notebooks. Diary in Canada. 3 vols. edited by William White, New York University Press, 1977.

Notebooks and Unpublished Prose Manuscripts. Edward F Grier, New York University Press, 1984.

Note: The title of this paper is taken from Wm O'Connor's book of 1866. Unfortunately I have not managed to get hold of a copy.

MAPPING KNOWLEDGE.

John Bently.

Three exhibitions that explore the idea of The Book.
Curated by Les Bicknell at the Minories Gallery, Colchester.
28 November - 15 January 1995
My adventure:
An artist reviews an exhibition
curated by an artist.

Today I am sitting in my studio. It is snowing outside. I have just made a book. I have made it in a way that I have never made a book before. How did this happen?

Let us go back twenty-four hours. I am lately arrived at Colchester station, slightly bewildered. It is snowing. Jane Rolo, from Bookworks, and I (a chance meeting) share a taxi to the Minories. What is a Minorie? Is there somewhere close by, a Majorie? The taxi driver ('I used to be in showbiz') knew everything about Colchester, except this one detail.

Cold and ignorant I enter the gallery. Before me are what look like four white burial ships. Removed from Sutton Hoo, up the coast? What manner of East Saxon devilry is this? What relic-offerings do they hold?

Dare we pry ?

I'll be nine-tailed, ship-mates, here be treasure! Books, books and more books, each as different as a snowflake.

We have come to the right place.

These are my most coveted specimens:
Mike Horl's *Something to drop on the way home,* a cardboard 'H' in a box of straw. Witty and simple.

Sophie Artemis', *Burying the hatchet,* animal jawbone and teeth and other things attached at one end by string; a book turned into the bare bones of its meaning, literally.

Beneath this , Denise Hawrysio's, *Killing I-IV:* Zebra-skin pages, stripey fur the only text, evoking the animal that was, and has now become a poem, despite our protestations.

Chris Taylor's hanging x-ray plates got me thinking of the relationships between books and bodies (ribs curling out from a spine)...

...Here be monsters

Roger Brown has made a Giant's necklace of houses, each house roofed with a book. Perhaps the vast creature that wears this lurks somewhere near?

I survey the cave for signs of bones. I walk on past Ian Hamilton Finlay's book of carved stone, inscription half-obscured, up some polished but still creaking stairs. One last glance. I see Ron King's dancing 'Acrobat' and Paul Johnson's psychedelic dancing pop-out wonderland being menaced by a Liliane Lijn 'Poem Machine '. At the top of the stairs I see an altar. Offerings to the monster?

Here are Janet Beckwith's murdered books, hewn and sewn and left to rot, unread. Here are books bandaged and sewn shut and imprisoned.
'encased in glass, held in space with not much air'

The monster is blind, I surmise. The artist teases its hunger for more and more useless knowledge. Here are my books, the antithesis of an encyclopedia, she says. I see a bandaged heart. I see stitching corroded by the sea. I am reminded suddenly of the Titanic, of objects dredged up from the ocean floor after a hundred years. Books unopenable but by no means unreadable. Iperuse my catalogue:

"Words worry me, they are such powerful and inadequate things."

I am moved by this and also this:

"I use books as vessels; they contain and hold safe memories."

Les Bicknell, nude but for his curator's hat, has, in keeping with a format often found in his work, created a triptych of an exhibition.

Part One consists of sixty or so books exploring widely 'some stuff known as book-art'. Janet Beckwith and six others form the central panel, having been chosen for their exaggeratedly differing approaches to 'The Book' under the heading 'Visions of Bookness'.

It is here that we find Mark Pawson, Mail-Art Superstar, who has doodled with his rubber stamps all over the toilet-white walls of his gallery. At its epicentre is a book containing 'all my rubber-stamps'. The book is bound in pungent rubber.

Matthew Tyson has exhibited some paintings of circles. At the centre of these is a letterpress book containing text relating to the author's baby.

"The book is not just a physical object, the essential, definitive icon of our material culture. It is the pre-eminent medium for storage and delivery of linguistic creation and information."

And ends with:

"Some readers will welcome new ways of understanding themselves in the ghostly

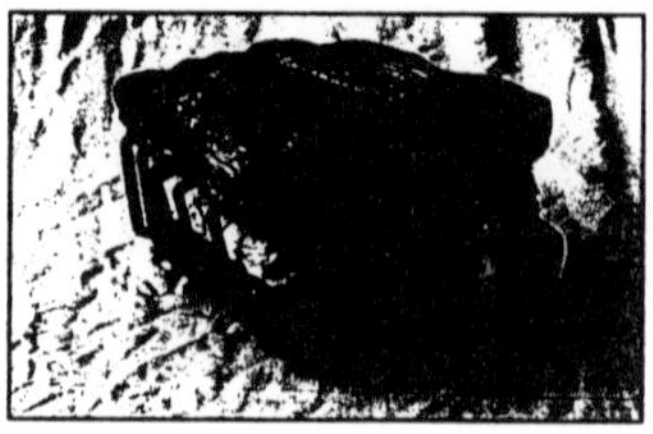

fragmentary, drifting leaves of an evolving metaphor.." The only image in the book is again a circle, caused by a sticky Ribena beaker. The book is bound in easy-to-wipe Formica.

Victoria Bean displays typographical constructions as deviously folded as ordnance survey maps.

An unpretentious table containing examples of all their collaborations, comfortingly dog-eared, announces the presence of Helen Douglas and Telfer Stokes.

I thumb through their newest book, *Water on the Border* . It contains Chinese poetry translated into vernacular Scots by Brian Holton. It also contains photographs and drawings linked by a cunning pictorial scaffold similar to a folding screen.

Like all Weproductions, the subtleties are too subtle to be adequately described using mere words. They have attempted to use the book-form as a device for connecting Chinese and Scottish culture, discovering a common borderland between the two:

"Chinese books and bamboo construction provided the inspiration for this device, which in its metered rhythm recreates in book-form our experience of Su Causeway, on West Lake, Hangzhou."

Panel three of this floating apparition of an exhibition is called 'Footprints in the Sand'. Suddenly and magically, the curator shifts shape and is once more a mere artist, a whole room of his work.

At first glance, I muse, what we have here is a formal exploration of the plastic inevitabilities of The Book. Chunky wooden covers laid atop piles of hand-made paper. Often in threes, sometimes a gold edge, sometimes stitching. I am reminded of the current debate about 'meaning' and 'reading' in post-modern art. Here is the debate playfully enacted, a celebration of pure (ish) form utilising the very 'Icon of our physical culture' and repository of all things, The Book.

Here are the still-solid husks of books sucked dry of their literaryness (by John Cayley's Infernal Machine?) Is this not story enough?. He uses the classic Triptych form, but also oddly empties it of its religious altarpiece connotations.

It is as if the artist, after too long in curatorial purgatory, having, with much hair-pulling, selected only those artists necessary for the defining of a new 'Theory of Book-Art', suddenly rebels, tears off his constrictive hat and creates a series of work to mischievously defy that very theory.

(And just when you think he's serious, you spot the piece of work entitled 'The Offside Rule Exception'.)

Whatever. These are works of art created in the same way that the works in the rest of the show have been chosen, with sensitivity, humour and restraint.

The raw materials are delicately balanced, chosen for their colour, their texture and their weight. In this same way are the elements of all poetry considered.

As I think this last thought, silence descends on the gallery. Outside, the snow has stopped falling. I hold up the cover of a book. Scorched and blackened. I hold it up to the window, it contrasts starkly with the pale sunlight. It has acquired a curious halo of snow and wind. I hear the cackle of ancient flames as I hold it up to my ear, whole libraries heaped up and torched.

Before I leave the show, I make my way into the first room; the first panel of the triptych. Here again are the wooden burial ships containing all that remains of our dying culture. I want to tear them open and steal their contents. I want to devour them with my eyes, my mouth, my fingers.

I am back in my studio. I have in my hands a pair of scissors. I am about to make a book in a way that I have never made a book before.

MULTIPLES AND MANTRAS

Louise Neaderland

Just as certain religions seek spiritual enlightenment through the repetition of a single, sacred syllable, a mantra believed to embody the divinity and to possess magical power, the visual artist can, through the repetition of a single eloquent image, create a narrative bookwork which explores and expands upon the original image. While it is the human voice which chants the mantra, it is the photo-copy machine which, in generating copies as it were, vocalizes the mantric image. In each case, the choice of mantra or image is critical. The copier can free the artist to meditate upon a single, vital, visual moment selected from the storm of visual stimuli generated and disseminated by modern technology and communicators.

The analogy of chanted mantra to repetitions of a single image is only partially valid. The mantra is itself unaltered through iteration, the way we perceive an image is itself altered by its repetition. The faceless, hypnotic droning of the mantric syllable frees the meditator from temporal restraints and allows him to walk into other spaces. The artist, a more active meditator, bent on creating and communicating with a viewer, manipulates the repeated image to create a narrative, extended in both place and time. Images are overlapped, cropped, enlarged, reduced, lightened, darkened to focus our attention upon a particular attribute of the original image.

The four xerographic bookworks to be described and illustrated in this article make use of all the above techniques. These books have in common a concern with timeless human issues and rely upon the symbols of sand, water, speed and fire to make some comment on the human condition.

Sadat's Journey: B&W xerox, 8 1/2"x 91/4" accordion folded. Opens to 81/2 x 56", 1982. (Sand.) This book is based upon a photograph by David Hume Kennerly of Anwar Sadat which was widely published following his assassination. Parts of a N.Y. Times text have been collaged along with the photograph, to document a particular time and event different from the time of the construction of the pyramids yet so much the same in terms of human conflicts. The photograph, evocative and dramatic, depicts a modern politician before the tombs of ancient rulers in an ancient land. It memorializes one man whose journey "began in the army and was ended in the army"[1] The method used to create a narrative from this material involved the use of twelve copies of the original image and text. The desert extends through the first three sections; a pyramid emerges as Sadat begins to appear in sections four through seven; we arrive at a copy of the original photograph in the eighth section, while the figure of Sadat, cut from the following section falls from the picture plane into the text, leaving the shadow or memory of the fallen figure standing before the pyramids (for one brief frame) then gone, leaving the pyramids disintegrating until only the sand remains. We have come full circle, "For dust thou art, and unto dust shalt thou return".[2] The collaged text highlights the particular man and event...ADAT DARING ARAB PIONEER OF PEACE..YPT: HE DARED TO DO THE UNTHINKABLE..ASSASSINATION IN EGYPT and the overlapped type suggest Arabic characters. Thus, 12 xeroxed copies of one image have been used by the artist to comment upon, and perhaps expand, the content of the original photograph.

Scenic Tunnels: B&W xerox, 14pp. stapled, 63/4" x 81/2". (Water) This book, one of a series on ageing, was inspired by a photograph taken by the artist at Niagara Falls and another taken during a visit with her parents at a nursing home. The title is taken from the Niagara Falls attraction pictured in the photograph and applied, with irony, to the corridor of the nursing home which terminates in the watery reflection of lights upon the highly polished floor leading to the French doors clearly marked EXIT, in accordance with the fire regulations within the home.

Not pictured in the illustrations from the bookwork is the preface, re-printed from the nursing home brochure:

FACILITY INFORMATION

THE BUILDING HAS BEEN CONSTRUCTED TO MEET ALL THE LATEST APPLICABLE BUILDING CODES AND FIRE STANDARDS. ALL ROOMS INCLUDE A WINDOW WITH A VIEW, AMPLE CLOSET SPACE, BED, BEDSIDE CHEST OF DRAWERS, MIRROR, WALL LIGHT, CHAIR, NURSES CALL AT EACH BED AND SPACE FOR PERSONAL FURNISHINGS. PRIVACY CURTAINS ARE IN ALL SEMI - PRIVATE ROOMS. OUTSIDE PATIOS AND WALKWAYS ARE CONVENIENTLY LOCATED.

This book is about the wheel-chair bound resident of the home and her journey down the corridor, that "Scenic Tunnel" which terminates with the watery reflection at the exit door. The method here is somewhat different from that used in *Sadat's Journey.*

The original photograph is successively enlarged and darkened; the woman vanishes in the darkness as the door is brought closer and made larger. Our attention is focused on the doors and the shimmering reflection of light on the floor, the doors and reflections are multiplied, the doors drop away leaving only the watery reflection and the last image is one of the ocean. Although three photographs have been used in the book, only one, that of the ageing woman, has been manipulated to eliminate elements of the picture as they become progressively irrelevant to the journey. We are left with the image of the ocean from which all life emerged and to which, in this case, it returns. Both the darkroom and the copier have been used in the creation of this bookwork.

Empress Bullet: B&W xerox, 8 1/2 "x 91/2" accordion folded, opens to 76". Title silk-screened in red on the cover. Off-set edition issued by the Women's Studio Workshop, 1982. (Speed). This book was created from a photograph by Vic De Lucia and an article written by Steven Crist (c N.Y. Times, 1982). Here, the copy-camera and the photo-copier are used to enlarge the already half-toned news photo, multiples are collaged, sections of the article have been excised and set in press type and the entire treatment of both the words and images contributes to the sense of urgency and speed which is at the heart of the narrative.

What appears to be a routine racing photograph, is transformed when our attention is called to the riderless horse running last in the race. The title of the article accompanying the photograph leads us through a tale of great drama and pathos, touchingly developed by Steven Crist. Empress Bullet has thrown her rider at the beginning of the race; she goes on to finish first but, being riderless, is automatically disqualified from winning the race. In panic, pursued by an outrider attempting to control her, she impales herself on the track railing, dying a terrible death. Both *Scenic Tunnels* and *Sadat's Journey* deal with the slow, inexorable passage of time, *Empress Bullet* bursts upon us in a flash of repeated images, text and word play.

Over 30 copies of the original image are used to construct, or re-construct the "Death of a Thoroughbred at Aqueduct". *Empress Bullet* opens with a straight copy of the photograph and article as they appeared in the N.Y. Times. The first seven sections (copies) of the image are overlapped to show the gradual emergence of the riderless horse, while a block of text describes the action and, from parallel lines of the collaged heading of the article, the word DEATH emerges letter by letter. Three images of Empress Bullet, seeming to race against herself, surge past the field of horses; four more frames of the riderless mare carry us towards the finish line as the story uncoils and words like "horoughbred, roughbred, troughbred and dred red

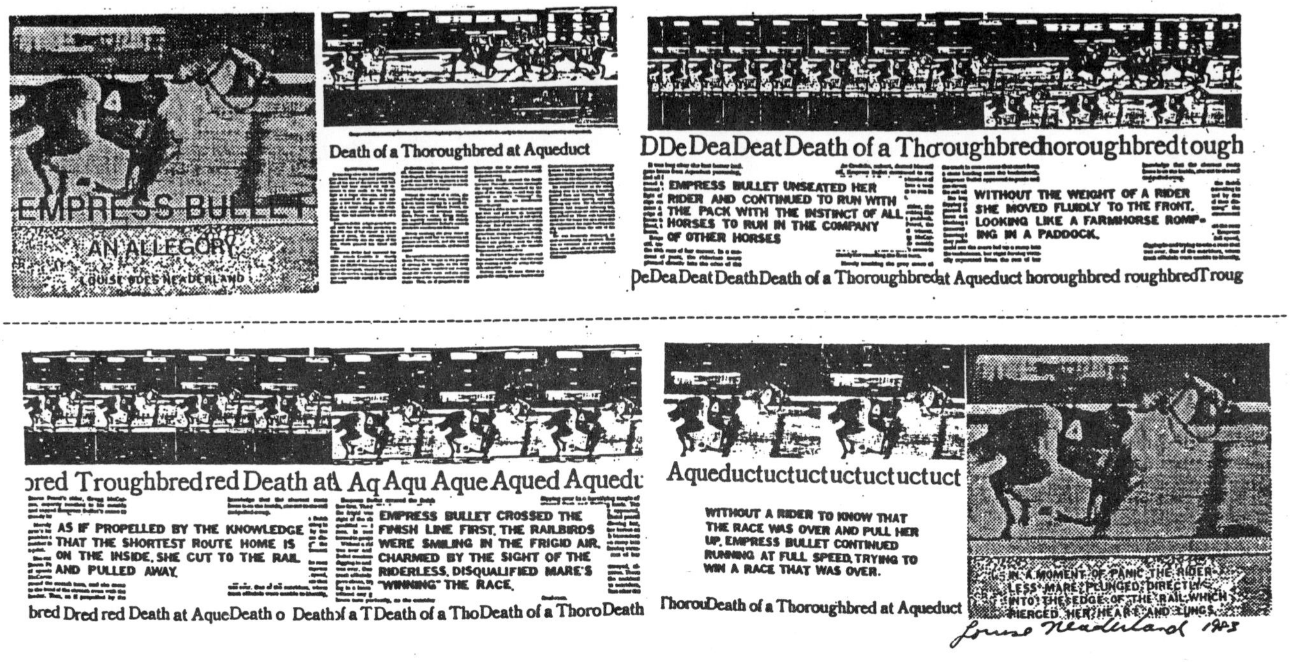

Empress Bullet, accordion fold xerox bookwork by Louise Neaderland.

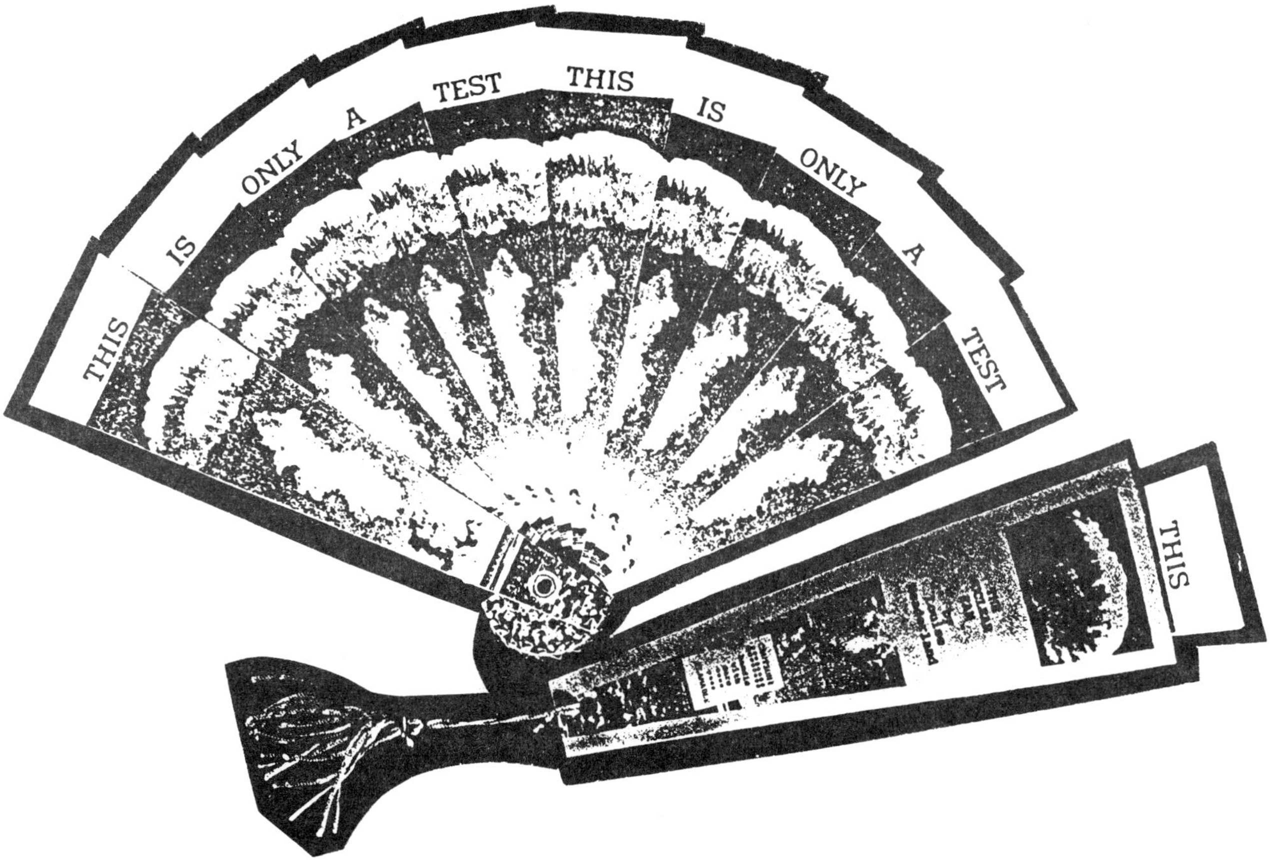

The Nuclear Fan , xerox fan with slip case by Louise Neaderland

death " emerge from the collaged headline racing along below the images. As she crosses the finish line , the half toned image of the horse is enlarged (for 3 frames), enlarged again, (for 2 frames) and finally a single blown up image, its component dots seeming to explode, carries Empress Bullet to her death at Aqueduct, uct,uct,uct. Who is the winner, the rider or the horse? If you run the course at your best speed to the finish line have you won the race? Who knows when to stop?

The Nuclear Fan: B&W on pink card stock, 10 grommeted segments open to a fan; red slip case with tassel, 1984. (Fire.) This book depends, in large measure, upon its form to convey its simply constructed message...Fan Yourself With A Nuclear Breeze, THIS IS ONLY A TEST. Ten copies of a segment of a photograph of a nuclear blast have been fastened together, only the word on each of the five segments changes and the change is repeated on the next five segments sending the message THIS IS ONLY A TEST THIS IS ONLY A TEST. A notice on the slip case warns us "Caution, handle with care, flammable!" The fan form itself recalls the culture of the first city to be devastated by the atomic bomb. The gentle symbol of the fan stands in dark contrast to the message it bears here of fire and destruction.

In each of the books described here, the artist selected or created the image (mantra), the copier repeated it and the artist synthesized or created a visual meditation upon the original image. This whole, greater than the sum of the parts owes its inspiration, facilitation and implementation to the ubiquitous office copier, most often thought of as an instrument of deadening repetition. Unlike Sonia Sheridan who appears to scorn this basic attribute of the copier and speaks of "its real capabilities, beyond copying"[3] book artists can use copying as the peculiarly liberating tool it can be, freeing the artist to explore, through the repetition and alteration of one image, a variety of philosophical and visual issues. In the parlance of our contemporary fast food, fast sound, fast image, fast feeling culture there are buzz words and bell words. The former, empty repetitious, yielding neither growth nor development call us to brief, sharp attention. Bell words and images, with time and attention may resonate through our lives and, lead us to great depths of feeling and understanding.

The concern of the artist should be philosophical as well as aesthetic. One means of dealing with universal issues and basic elements is to look at fewer things more deeply. The technique for creating bookworks described here uses a high speed copying technology to slow down and refine the artist's perception and to animate the heart of a single image.

1. "Sadat's Journey", an article by Fouad Ayami, N.Y. Times Oct 7 1981.
2 Genesis 3:19
3 Sonia Landy Sheridan, "Generative Systems Versus Copy Art A Clarification of Terms and Ideas", *Leonardo,* Vol.16, No 2, pp 103-108

Tagore Nagar by Amanda Burrow

Pondicherry,India: Dupius Press, 1992
available from Printed Matter, New York
(1994 - $15)

A rosebud left between the pages of a book acquires a curious two-dimensionality. The third dimension seems to have been sucked out of it. The whorls of the petals can be seen to have been circular, the leaves can be seen to have projected in all directions, but the fossilising pressure of the book has reduced the rosebud to a diagram of itself. And yet, one cannot help think that a breath, a kiss, might restore the missing third dimension.

Tagore Nagar is redolent with the sense of an immanent third dimension, indeed, since it seems to be intrinsically about a fourth dimension, it mutely insists that a third dimension must permeate its wafered plane.

Each of the two hundred copies of this book was cut, assembled and handled by women in India. Their hands shaped a shrinking tunnel of space that recedes into the heart of the book. Having reached its centre, their fingertips then articulated intricate mechanisms of inner space before bringing the book to a close.

There is a small hole in the front cover, though like most of the shapes and pages in the book it is roughly delineated in black. You might at first, therefore, perceive this hole to be a solid, an orb, especially as it is positioned over two enigmatic red lines that read archetypally as horizon. Venturing through this negative sphere to the interior, you are set spinning in a layered space. Successive pages have diminishing circles cut into them. Various of your organs, especially the eyes, ease through microspace ever deeper into the body of the book.

As you turn the pages you lose your bearings and float in the limitless space of the page. Space exists where you might expect to find solid. And you tunnel into solids that have metamorphosed into voids. Glancing back, you see a circular shutter poised above the previous opening as if to suggest that you are actually in a submarine or in a spacecraft viewing the route already traversed. One stop past this point, in the very centre of the book, you definitely feel inside, not outside. Having voyaged through space in your skin-tight body you

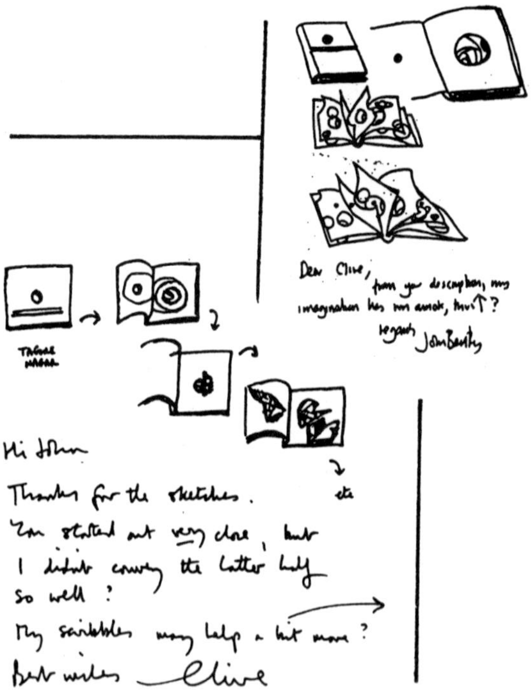

have arrived inside yourself. But your perceptions oscillate, and it is still possible to imagine that your self-craft is venturing through galaxies or nudging unicellular organisms. Vast perspectives open up.

There is light ahead. Through an inviting slit another odyssey beckons. Turning the page, penetrating further into the space, an incoming object is revealed to be a complicated spatial structure. Pushing though another layer you glimpse the spokes of inner space stations. These absent volumes evoke substantial bodies poised silently and virtually unmoving in a vast space. You are privileged to see successive views of the compressed, but unfathomable, structure of a polymorphous universe, whose size might be measured in light years, or microns.

The machinery of the spheres is sliced as each page turns. But then another shutter is poised to close. Turning again, the overlay shrinks the celestial mechanism to a planar armillary sphere; your insight is diminished. And yet, as you withdraw from the book, you can still look back and glimpse the girders of the silent spheres of the engine of the universe.

Feeling your way through the virtually two-dimensional body of this book, almost stroking the elusive spheres with your fingers, you get an acute sense of the missing third dimension, as if, given the right conditions, the flat space of the book might suddenly engorge, suck you in, and whirl you into unimaginable depths.

Clive Phillpot

MEDIATED IDENTITIES:
WOMEN ARTISTS' BOOKS IN THE 1970s
JOHANNA DRUCKER

Artists' books achieved a currency in the 1970s mainstream artworld which they had not enjoyed since the early avant-garde experiments among Russian Futurists in the 1910s. The book form suited the agenda of '70s artists seeking a way to circumvent the institutional power nexus of the artworld and put work directly into circulation in the form of inexpensive, mass produced, multiples. Available reproduction technology in the form of offset and xerox (as well as letterpress, though that largely remained confined to the fine print, limited edition ghetto) combined with aesthetic practices intent on defining themselves in antithesis to the tradition of the fine art, high art, autocratic object, fostered a boom in the investigation of books as an artist's medium.

The '70s was also the decade in which the women's art movement gained significant momentum. While many of the books discussed here come out of the American scene and reflect the varied concerns of women artists in the USA, similar impetus for expressions of feminist concerns could be traced among women artists in Europe, Canada, Mexico and elsewhere. Mine is a somewhat random sample, culled from an exhibition curated by Laurie McEnery at Printed Matter in New York in 1993, This selection doesn't adequately represent the exhibition's breadth and depth: McEnery, working from the on-shelf inventory of Printed Matter, unearthed a poignant record of intense productivity which has received hardly any critical or scholarly recognition. However, on the basis of this selection, I want to examine two things: the aesthetic and conceptual issues expressed in this work as evidence of the concerns of women artists working in that first wave of highly visible feminism, and the way in which books as an artistic form facilitated and aptly suited the need to articulate and circulate these concerns.

One of the hallmarks of the women's art movement was the foregrounding of attention to women's identity as an aspect of their art production, as with Judy Chicago, Carolee Schneeman or Eleanor Antin, for instance. This trend is evident here in a variety of ways. The presentation of the artist's body, everyday life and/or personal dilemmas appear both as thematic and formal aspects of the work. Martha Wilson's *Autobiography* (1979), is paradigmatic of the intersection of the women's movement, art world, and artists' books. Both in terms of production and thematics, Wilson's work is a classic artist's book of this genre. It is small in format, composed of a single signature to make a 16 page book which was obviously printed as a single, larger sheet. This kind of design, efficient and economical, conceived from within the specifications of a printer's standard paper sizes, with a minimal staple binding through the spine, was engineered to take advantage of the limitations of the production process. In this way, maximum use was made of each print run (each pass of the paper through the press) and a book was produced with a minimum of means (especially money). This book was produced by Conrad Glever, publisher and printer of Chicago Press books — an enterprise devoted almost exclusively to artists' books, which means that no commercial printer had to be involved in the process (except perhaps for making the color separations). The book thus demonstrates the non-commercial network of artists working in collaboration with each other to make books which exist outside of the frameworks of traditional publishing. The costs of production were probably not more than that of paper and film and printing plates.

Autobiography has two alternative stories in it, both of which present the artist within photographic narratives, thus allowing her to simultaneously enact and tell her stories. The first very simply recounts the routine of Wilson's day, documenting her rising and work pattern through to evening party or dining and back to sleep. This first section extends over three pages with black and white blurry photographs of Wilson — getting out of bed, working, on the phone — narrated with typewriter produced text in flat documentary style. This narrative, which breaks off with no finishing punctuation, is interrupted by a second which uses handwritten script, blurry (out of register but not necessarily deliberately so) color images, and a description of Wilson's "alternative" life in drag, in short hair, long hair, in crises of self-image and identity. These investigations finish off the remaining nine pages of the book and force the first narrative's flat tone to break open into the psychic arena of Wilson's interior life. The efficient (if blurry, out of focus) alternative arts administrator/artist of the first few pages is contradicted by the later image, which neither matches the first, nor explicitly negates it. The sense of self in the two parts of the work are equally posed, equally about posing, struggling to find oneself in postures, actions, the solidity of either activity or appearance, both of which, in fact, prove to be elusive here in spite of the obviously "autobiographical" tone. Wilson may not have intended to deconstruct the genre of self-recounting, but she does, and the limited space of the book's format compresses the disjunction into a felt relation: one has to read the two narratives against each other because of their sequential proximity.

Autobiography is not a great work — but it does serve as an excellent example of the way the format and wide open aesthetic parameters of the artist's book were suited to individual expression. This is not a slick work, not a contained and polished and produced work. It is a fast and sketchy presentation of a particular moment, point of view, and "take" on the condition of living as a specific woman in the artworld in a particular moment in time. Its very specificity is what allows it to be generalized, so that one can readily extrapolate from Wilson's experience into one's own, or that of other women whose lives center around the business and practice of alternative art. As a document from that life, in its various conflicts and contradictions, *Autobiography* offers insight and specifics. Again, this comes in part from Wilson's willingness to use her self as her model, a gesture which is a curious combination of narcissism and self-loathing, a bald self-confrontation and a partial self-evasion - her poses are all poses, none therefore more "authentic" than the rest. There is not a control image against which the rest are read as variant, all the images are variants, versions, which present a mediated construction of identity without fixing it. Whatever Wilson's intentions, and they may have been far more forthright and direct, the effect is this complex of images which continually displace each other as "her" throughout the book's pages. And the book, keeping them together, forces that refusal of closed meaning into circulation so that the bright red, purely typographic, cover which announces itself as an "autobiography" trumpets its specious authenticity in the unhyphenated break of the word into two non-standard parts: "autobi" "ography."

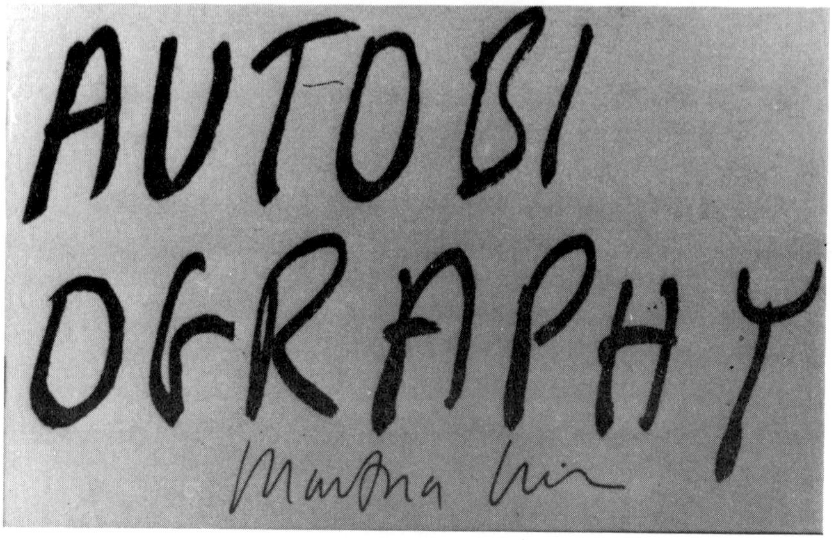

A.Martha Wilson, *Autobiography*, Chicago Books, Chicago, 1979 (cover)

It is an almost obvious trope of the book form that the sequence of pages can mimic the chronological passage of time. In Barbara Schmidt-Heins' book, *1949-1979,* attention to her identity is restricted to the single image of her face from infancy to adulthood. The transformation from image to image is sufficiently dramatic to contest the assumed continuity of self, and there is a definite tension between the sense of a kernel identity at the core of these images and the continual reinvention of one's self according to circumstance, situation, and, of course, according to the gaze projected onto the face which answers back as the image photographed. Identity does not resolve here either, and the issue is not so much one of trying to select or even synthesize a single image out of this regular sequence, as it is of trying to sustain the heterogeneous and polymorphous nature of identity itself. The images are small, identity-card sized, and printed on dull coated stock so that they read clearly, finely, out of their wide white margins. The book is clean, but not clinical, and, as a small gallery, functions with reassuring neatness to contain its full measure of contradictions. The "what is a woman?" question of the '70s combines with the always compelling "who am I?" here offered up as a kind of nostalgic "who was she?" There is no particular irony in this work, and it is the very dullness, the degree-zero, uninflected quality of its presentation, the documentary format, which allows the work to offer itself for so many possible readings.

Two books from 1979 by dancer and performance artist Suzanne Lacy explore two different sets of clichés: personal and professional. The first of these, *Three Love Stories* literalizes the corporeal metaphors of romance. "A Gothic Love Story" opens with a black and white photo of a preserved heart in a jar being passed from a woman's hands into a man's grasp. The photo is captioned, "She gave him her heart." The heart is enormous, obviously not human, but becomes the joke image around which the grotesqueness of the story revolves. Neither character is ever fully pictured, and the jar with the heart is the epitomized, condensed image through which the course of the romance is traced. In the second story Lacy herself is pictured with a set of huge lungs, and her grimacing interaction with these organs (also too large to be human) provide the visual element of the narration. The third story, titled "Under my skin: A pornographic novel" pictures masses of organs, mainly intestinal parts, in clean bedsheets, accompanied by the standard language of romance novels.

The self-conscious investigation of the roles and terms of romantic love were a prevalent feature of the women's movement, explicit in its literature from the first writings of Betty Friedan. Such investigations have their precedents in the work of Simone de Beauvoir and Colette, Jean Rhys and Virginia Woolf, among others of course, but the 1970s expression of these concerns in the visual arts also incorporates the engagement of fine art with pop and media generated images, not merely conventions of high art and literature. This move towards the realization that identity, subjectivity, and the received narratives according to which women shaped the lives they actually lived, were all

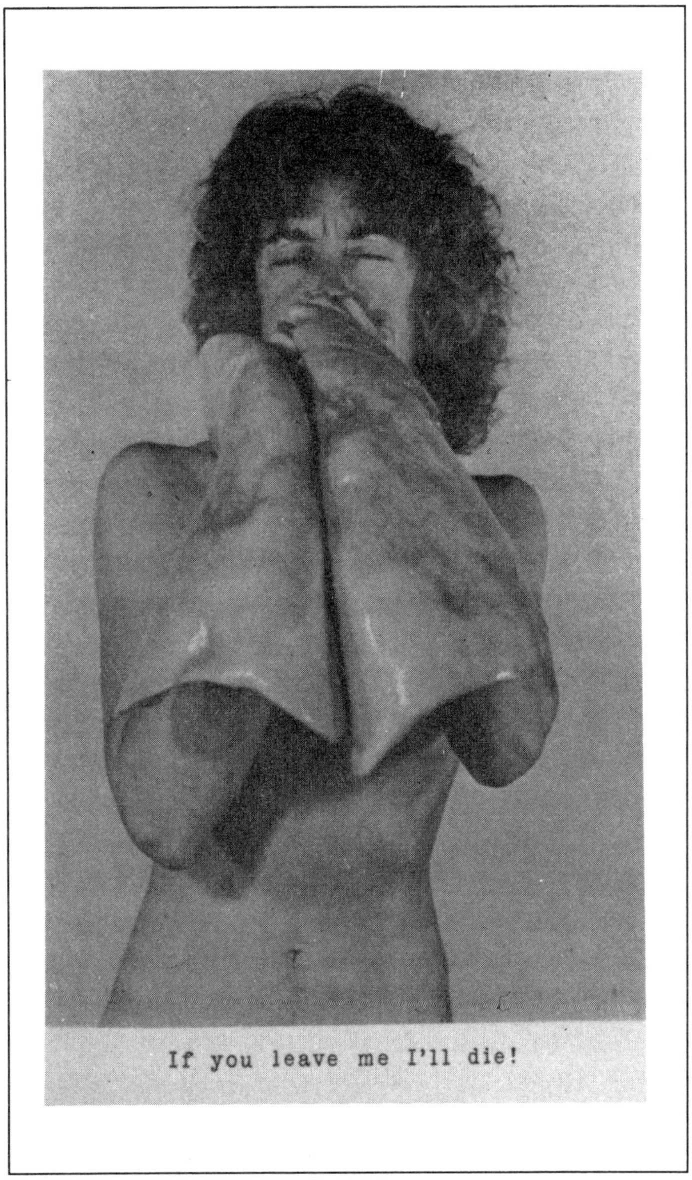

Suzanne Lacy, *Three Love Stories*, 1978 (no place or publisher indicated)

produced in the broadest sense by the culture industry, as well as by social mores and conventions, widened the vocabulary of the visual arts. The 1970s witnessed an increase in appropriation as a strategy for critique of those very mainstream sources from which the images are taken.

Lacy's visuals work in marked contrast to the language of her narratives whose source in pulp fiction is apparent in their lurid expressivity: "take me, she moaned." But who, then, is Lacy in this? The presence of her body, naked, if discretly covered by the large, slimy looking lungs in the second story, clearly marks her as the speaker of these tales. At the same time, the ironic combination of image and text keeps her distinct from the woman of the narrative voice. She both is and is not, simultaneously, the woman whose story she tells, both is and is not "told" herself as the woman of the tale. The dark humor of the work, and, again, its small format, accessible, intimate, immediate, and affordable, produced in an edition of 500 copies make this an exemplary book of this form. The book participates in the kind of democratization of artistic expression which was a prevalent ideal of the activist spirit of '70s feminism. Highly personal, idiosyncratic in tone and sensibility, and unlikely to find backing in the commercial publishing world, this is a work which required relatively little financial support for production.

The second Lacy piece is slightly more ambitious in terms of production values. Full color throughout, the work is small enough in format to still have been affordable within a relatively limited budget. Titled *Travels with Mona*, the "book" charts its tale in parodic imitation of a tourist's postcard portfolio. The images fold out to document the travels of a woman artist and her canvas, a lovely half-finished paint-by-numbers-ish version of the Mona Lisa, through the major sites of a European grand tour. From Basel to Milan to the Swiss Alps, we follow the young painter in her quest for "visibility, acknowledgement, fortune, even immortality." The second side of the card file moves the artist into the exotic landscapes of British Honduras, Tikal, and finally, back to Los Angeles. The investigation of women artists' identities in relation to the history of the European canon, an on-going issue today but a newly charged one in the '70s, serves as the basis of this work. There is no other "persona" revealed here besides the unidimensional image of "a woman performance artist" who uses the canvas of the Mona Lisa as a point of reference against which her own historical position can be evaluated. The identity she has, is therefore mediated several times, through the role she takes on, the historical referent is itself brought up for question. The image of the artist, for instance, in the midst of a wide expanse of dry Texas semi-desert landscape dwarfs the Mona Lisa in relation to the immense sky. If the question of women artists' stature in relation to the canon is one part of this equation, the relevance of the canon to the indifferent terrain of the American southwest introduces another set of considerations of scale, pertinence, and historical and cultural relativism. The question for women artists, "what am I in relation to that canon?" becomes inflected by the second question, "what is that canon here and now?" The printing of this project is slightly blurry, the typographic format of the cover not quite up to the graphic standard of even the most banal of turnpike gift shop fare, and this, again, marks the work as of its own idiom and location. The piece passes, at first, for the slick, commercial form it imitates, but then is subtly distinguished from that format by the small differences in production, the large difference in sensibility.

Not every investigation of identity by women artists in the '70s focused on themselves, though the compelling urgency of defining professional and personal identities according to feminist terms was an overriding concern. The pieces in Angelika Bliese's *Who's Who* (which includes two works) all focus on images of men. The first "Blauaüglein" contrasts two group photos of German officers, possibly in the Luftwaffe, with their hats on and off. Their heads, vignetted, blown up, isolated, and placed on facing pages allows them to be seen as officers and as men, in and out of a role which stigmatizes them historically, and opposed to the more neutral presentation of physiognomy unmarked by cultural signs. The second part of this piece documents masculine endorsers of the paper, *Die Welt*, each of whom have given their testimonial on its front page in tiny features titled: "Why I read *The World*." Also excised, reduced to their heads, these men are printed in black and white grainy

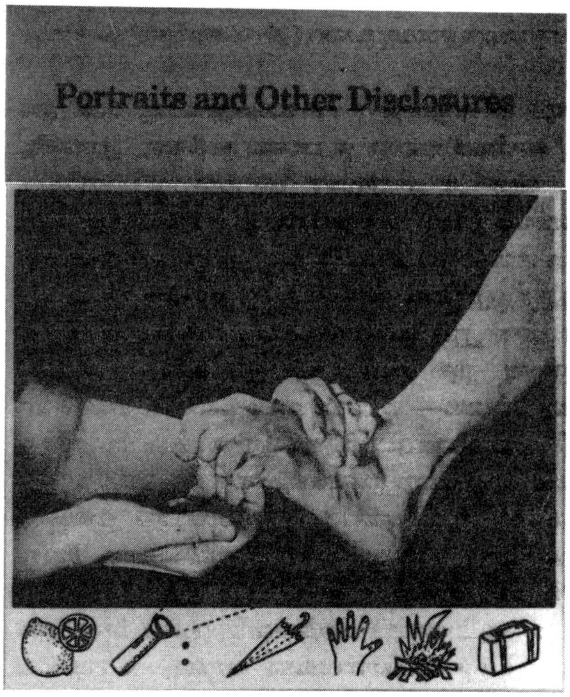

Gwen Widmer (maybe), *Portraits and other Disclosures,* Monticello, Iowa.

blow ups which emphasize the half-tone dot pattern of the newspaper source for the images. A bright blue solid fills their eyes, in emphatic repetition of the Aryan genetic preference. Who are these men? To an American, myself, in the mid-1990s, their faces resonate with no more specific identity than do the faces from a Christian Boltanski monument of random high school children. These are not the historic faces of Hitler or Goebbels or Goering, these are men whose activities and roles were such that they are increasingly obscured by history outside of the realm of a localized memory. Historical identity diminishes quickly, recognition as well, and the peculiarly pop treatment of the images, appropriated, manipulated, represented in all the marks of reproduction technology included in the image, also situates them in the realm of the already depicted, rather than the made, drawn, examined. The "Who's Who" of the title takes identity out of a personal frame and into a cultural examination, a field of faces, images, and memories too vast to return to the personal as a single point of reference. This work has some of the rhetorical markings of the work of John Baldessari, and opens the parameters of inquiry into identity in a public domain.

The second piece in the book, "Selbsportraits," is a series of photos of a man's face, whose? We don't know, only that he is always on the right side of the page, as one half of a pair of images which map a continuous landscape across the gutter of the book. The left side is empty, and the "self" of the portraits elusive. Who is "he" that in a work by Angelika Bliese he returns her gaze? The "other" who is the lover? Is he in fact the photographer of these images, holding the camera at arm's length, letting "her" be the complementary empty image to his left? Or is she the "self" defined by his returned look in the intimate exchange of companionship and love? Thematically, however, this work extends the vocabulary of identity as mediation — through photographs, back into print, and into print again, always with self-conscious attention to this circulation of images through venues and sites, and through the complex interrelations of an intimate gaze returned and inscribed.

Elusiveness, rather than lack of resolution, characterizes Gail Widmer's *Portraits and Other Disclosures*. Evidently a highly coded personal memoir, the book opens with an account of a group photo taken on a very hot day, a "pile of flesh" comprised of limbs of members of an extended "family" unit, "Gwen," "Susan," a narrator, the narrator's son. Three different diary entries face the photograph, a Weston like abstraction of form and textures. Each of the entries is entered in a different hand, or typewriter, and the presence of three voices is marked at the book's end with the statement, "It will be a salute to trigonometry." Here the personal frame of references is occluded behind the surface of the photos and statements, none of which give access to the events which provided the point of departure for the work. The back cover of the piece contains a small photograph of the back of an angular young blonde woman standing in a country setting holding up a black and white photograph of a dirigible in dry dock. The captions reads, "Gwen Widmer: She Lives in Cedar Falls, Iowa, and This Isn't Really Her."

If Widmer's book (if it even *is* Widmer's) remains obscure, unavailable except as a playfully personal, idiosyncratic "disclosure" which only reads to the initiated of her group, social set, and intimates, then it also indicates another aspect of book production as an artistic mode — its capacity to encapsulate those obscure moments of communication. This book also offers an examination of romance, this time with peculiar attention to parts of the body. An ear, for instance, is presented in full page photo with the caption: "Dr. XX. He whispered sweet magic in her ear." Facing this page, an account of an operation on the tympan and middle ear, followed by a small, handwritten comment expressing pleasure at having been of assistance to the operation. Nothing adds up here, the parts are parts, the clues all tangential. The tone is diaristic, but not revealing; personal, but not explicit; suggestive, but of what? A portfolio of photos, a small gallery show of the same images, would have functioned similarly, but would not have maintained itself across time, as this does, arriving a decade and a half later like the flotsam and jetsam of some abandoned psychic life, still intact, but nearly meaningless outside its original

Linda Neaman, *Sex and Monsters,* Hallwalls, Buffalo, New York (no date)

frames of reference.

Martha Hawley takes these very notions as the premises for her book, *Notebooks.* This is the most elaborate literary production, the only explicitly literary production, among these works. Hawley's project, produced as a small, perfect-bound, pocket-sized paperback, mimics the format of the pulp detective novel, only to systematically deconstruct its conventions. Highly informed by theoretical writing, particularly French semiotics and psychoanalysis, Hawley's structure is designed to weave the illusion of relations among the characters and elements of the story as if it were, in fact, an actual mystery. No narrative unfolds, however. Instead, we are presented with a dossier of materials, interviews, notebooks, communications, transcriptions, all of which undo the possibility of returning to the scene of the crime, or its main protagonists. Each new bit of information leads further from that scene, and the whole is a continual dissolution of the possibility of narrative resolution. There is a self-consciously ironic feminist twist to the back cover statement "What Do Detectives Want?" which paraphrases the endlessly repeated psychoanalytic inquiry into the nature of women's desire, and sets up the frame within which Hawley's strategies are played out. This is a peculiarly historic work, one which struggles to fulfil and extend the critical issues of narratology and textual analysis which dominated the French theoretical scene in the 1970s. The impetus to write from that critical field as a point of departure is evident throughout Hawley's book. Though the quality of the writing, conception, and story are somewhat naïve and the characters clichéd, this brings *Notebooks* closer to the feeling of Dennis Wheatley's novels than to the elegant sleight-of-hand of a Marguerite Duras or Alain Robbe-Grillet novel. All identities are textual productions in Hawley's work, made only through the mediating structures of language and interpretative reading. What is striking, however, about this work in relation to, for instance, that of Lacy of Bleise, is the way in which the unquestioned essentialism so generally associated with the women's movement is undermined equally in these works, all of which assert identity as a construction, rather than a given.

While Hawley flirts with clichés, rather than fully embracing them, the work of Linda Neaman,

Sex and Monsters, aggressively attacks the myths and fears associated with women's bodies, procreative power, and sexuality. Laid out like a combination workbook, hygiene text, and clinical sex manual, the piece combines goddess mythology, pro-prostitution activism, and imagery reminiscent of the sex industry. Neaman's work, like that of Annie Sprinkle, seems designed to open up the territory of sexuality, remove it from its taboo spaces, and bring it into the more available domain of frank discussion. This is, again, indicative of the mood of the '70s, in which women functioned as an above-ground network and community for each other.

The "monsters" of the title are both men and women in this work, the fantastic creatures of projected fears and anxieties which trouble, charge, and complicate erotic psychic life. And Neaman's feminism does not pass judgement on the male or men, as much as reveal the complex structures in which sexual identity comes into play given the history of heterosexual relations in patriarchal society.

Heterosexuality dominates these works, and in that sense the anxieties of women artists of the '70s clearly circulated around the troubling issues of the price to be paid for independence, strength, work and professional activity. There are, I am sure, books which manifest the concerns of the lesbian community, but they did not show up in this small sample. Instead, it is the scenarios of the heterosexual couple which make themselves so evident, everywhere explored in personal, mythic, clichéd terms. Gail Rubini's *Forever Yours*, with its prose narrative and tiny color images of dolls and wax figures, also replays these concerns. The woman's point of view and her complicity with the mechanisms of her own repression are everywhere in evidence.

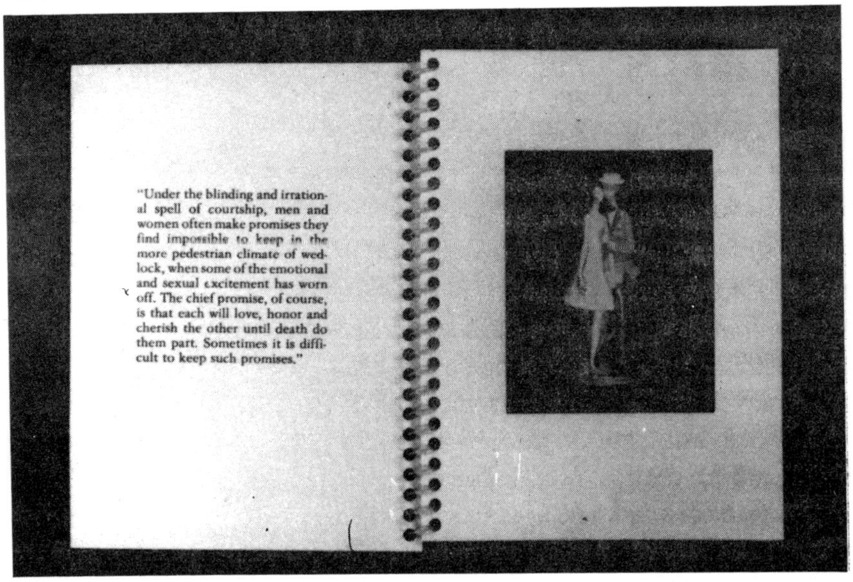

Gail Rubini, *Forever Yours*, Chicago Books, Chicago, 1980

This book is small, spiral bound, with beautiful little color images. The title is printed in blue script across the white, faceted plastic paper of the cover, and the marked "femininity" of the form of the work relates it to that of Laurie Simmons and other women artists of the late '70s who were reworking the clichés of mass-produced images of the "feminine" as a critique. Each opening of the book tells a different story of a different couple, and in each case, it is the woman's anxieties, issues, troubles, hopes and expectations, which map the relationship. As a mini-gallery of vignettes this piece explores the working out of identity as an aspect of the complex network of social and personal relations in which women lived and live. There is a clear sense that for these women (those of Rubini's stories), identity is formed in such relations, as much as in one's work or independent life, and that the traps, constraints, seductions of habit and social form are an integral feature of the internal life according to which choices are made, regulated, defeated, or achieved.Unique among these works is Alice Aycock's *Project Entitled "The Beginnings of a Complex...."* (1977). Aycock, a sculptor, made her reputation outside the realm of artists' books, and the book is as much a document about her working process (on a large, sculptural project) as it is a book for its own sake. Nonetheless, it succeeds on its own terms as a compact and rather exquisitely produced piece in which a pastiche of personal, appropriated, and working texts all interweave, combining as the conceptual field from which the sculptural complex — a network of underground maze tunnels built at

Documenta — emerges. That the complex was actually built, and is pictured here, hardly matters for the book. It might just as well chart a conceptual piece, so fully realized is the book object as a work. The pleasure of the work is in the slippage of Aycock's language from register to register, the synthetic imagination which takes the abandoned structure of a farmhouse on a median strip of turnpike and a hog farm, and a Giza pyramid, and forges from them a unique sculptural approach to construction. Aycock's presentation of these gleanings and gatherings transforms this work from a mere document into an autonomous piece, separating it from any dependence on a work in the world beyond the bound covers of the book. It is this capacity to function as a finite space, to contain its own references while alluding to — or eluding, occluding, confusing — external points whose existence need only be suggested, not proved, which makes the book such a unique form. It is independent, living a life of its own as a realized object, capable, unlike an exhibition or installation, of travelling across time and space intact and fully realized.

The sum of these books taken together — mostly work which would otherwise have been lost, in many cases produced by "minor" or little known figures — makes clear that for women artists of the mid to late '70s the artist's book offered a new venue for the production of concerns central to the identity of women as artists, the difficulties of managing relationships, exploring sexuality, one's place in culture and history and the history and practice of art. These themes all come through this body of work, all of which did not so much reflect as *make* the space in which such exploration could take place.

My own work started in that period, and *26 '76* (1976), *From A to Z* (1977) and *Surprise Party* (1977) were much concerned with the social construction of identity, and of the power of social conventions to contain and limit the exploration of one's self, work, and subjectivity. But what strikes me at this distance, aside from the pleasure of making manifested in all these books and the evident satisfaction of their achievement (for maker and, now, viewer), is the capacity of the book form to fulfil needs which could not be met by other art-world activities. The freedom of books from any controls or restraints (beyond meeting costs) meant that they were an unlegislated territory. In that period of the late '70s, galleries were still loathe to give solo shows to women, especially younger women, and critical reception was much marked by the position of the Women's Art Movement and the various forms of resistance or support it mustered. The range of work in these books, all of which are nuanced in their investigation of identity, all of which use the book as a mediating form in which to pose that investigation shows how much more heterogenous the reconceptualization of identity was in that period. Far from endorsing a literal or essential feminine position, these works almost problematize that position, imbricating identity in networks, social relations and conventions, rather than asserting its intactness as an *a priori* point of departure. There is nothing inherent in multiple pages and sequential ordering which insists on this heterogeneous and contingent conception, but there is at least the possibility of using these aspects of the book to permit multiple images and positions to come into play in relation to each other.

The last of these works is a tiny piece, a single scrap of paper, which folds down into a work as small as my thumb. Untitled, it bears the name Rachel Youdelman and the copyright symbol next to the date, 1976, on its left edge. Three black and white photos, one on each panel, depict a middle aged couple playing shuffleboard in a suburban back yard. The caption, which breaks up over the three panels, is as follows: "Whenever we kiss ... / I worry ... / and wonder ..." This is 1976, a pre-AIDS moment, and so the anxiety of intimacy is not located in the fear of epidemic contamination, but merely in the difficulty of relations. Who is speaking in this image? The woman is overprinted with magenta, and so seems to claim the subject position from which the piece is expressed. The piece is a kind of one-liner curiosity, hardly a masterwork, and yet in its tiny, condensed form it came into my hands as some kind of record of communication, some cry from the past. Who is Rachel Youdelman? Will I ever, never, know? And yet, one of her works, miniature, specific, peculiar, exists as a kind of quirky comment on the unfixity of things, identity, the status of objects, and the artist's book's place and places in the world.

WORD UNNECESSARY AS SUCH

Nikonova - Tarshis

Word's a malady
a sore
a fate
a ruin to poets
in the past and the present
and incurably
it pushes poetry
to death
to decomposition

Word is unfit to construct
the symbol of Poetry
A. Chicherin. "Kan-Fun" (declaration)

The problem of vacuum, which never occurred in former, so-called "general" literature, where visual and phonetic parts were not segregated, is presently considered in the context of other literary problems. It has been proved that SENSE depends not on LETTERS, or their arrangement and texture only, but also on the book surroundings, vector characteristics, manner of reading, page (platform) texture and the quality of other literary elements (or their substitutes). The 20th Century turned its most intent attention to vacant space, which is sacred, since it is always present in any text - in the text as such.

The platform of a page is not just a carrier of text; it is also a literary fact, like a house, which is a fact of human civilization, and not just an element of landscape. If the content of a book is the state of its platform potentialities, and nothing else (as in the multiple works by the German author Bernhard Meyer) such features as vacuums should not be treated as marginal. Besides, the page platform may have both phonetic (jingle, rustle) and tactile properties. The latter may be reached through the use of platforms of various texture: cellophane, iron, fur, emery, paper and even glass. Why should a "civilized" reader, when using a book, ignore the language of smells so common in the animal world? A vacuum page, if it emits a strong smell, is never empty. And why should a page platform be smooth only? A man of letters should make use of any surface in implementing his idea of a book. It is not by chance that some Italian authors (like Elio Garis) arranged their lengthy "pages" right in the air by hanging them on tree-tops.

"Overtone" platforms, "accordion" pages, hyperboloid platforms, (for example, the ancient form of book rolls which was developed in works by modern Soviet authors including the "Oduryaka" roll book by Sergei Sigei), and quantum books of page fragments. Somehow, the latter immediately recall Khlebnikov's pillow cases filled with manuscripts which the poet always kept with him on his trips in Russia. (Strangely, the glass "travelling-bags" of Meyer filled with newspaper fragments also recall those pillow-cases.)

In 1982, when I worked at the issues of TRANSPONANS, a handmade magazine on the problems of visual and action poetry (which was the only one such magazine in the USSR). I dreamt of making a magazine issue in memory of Velimir Khlebnikov, in the form of a pillow-case filled with page sheets. To my regret this project never materialised. But I was lucky to implement my project of a vacuum single-platform book-harp of discrete lines, of clean white paper strips fixed tightly on a wooden frame. During my performances in 1983 in the poetic shows of a transfuturist group in Leningrad (Nikonova, Sergei, Constrictor), by "playing" this book-harp with light movements of the hands with a razor, I obtained at the end of my recital, a completely new object: a book-frame with white scrolls of "line", fragments cut with the razor, hanging all over its edge.

The literary meaning of a clean platform gave rise to a completely new trend in modern vanguard culture, so-called "book-art", where the book is treated as an object, a sculpture, a form and where the "book" in the common sense of the word is only one of its aspects. There may exist such objects as the book as a whole (a monolith without pages). In recent years, such very showy pieces were presented in various publications and at expositions in Germany, Italy, Portugal, Mexico etc. (For example the works by Dario Tozzoli and Carla Bertola, Italy).

There is no end to platform varieties. They may be of border (with a hole in the centre) and disc-type (the disc-book by Luciano Cappelari, Italy). The page platforms, or the book in general, may take the form of revolving stands, cylinders, lanterns, or gramophone records. Of particular interest is the form of a book-record (with radical or spiral text arrangement), or of a lantern-book lit from inside. Turning the pages about the book as an axis, and communicating with the book by tearing out or burning its pages, or, on the contrary, by inserting your own texts and all kinds of supplements (a curious fact to remember: once Khlebnikov was reading a book which he lit by burning the pages he had already finished), we re-create the book as a kinetic object and the surrounding atmosphere of participation, if not in the creation, then at least in its destruction ("self-destructing" art comes from the same source). Besides, destruction, or deformation of the "normal" text composition (fragments of letters, pages, books, blank spaces in the text), also herald the coming vacuum.

Vacuum poetry is the poetry growing from a genealogical tree, which from the moment it bears upon a clean platform finds itself in close relationship with other arts and sciences fixed on platforms. Yet relationship with spatial arts, e.g. the theatre, is also possible. In the latter case, action, that is rotation, turning of pages, folding crumpling, breaking, pasting together, or making the book by trampling with one's feet on its pages (Sigei), comes to the forefront as a sole element of literary existence. In this situation, action is perceived as text, and the process of platform treatment , is the act of creation. The simplest everyday actions start to arise a specific interest. For example, closing a book the reader fulfils an action of text superimposing on the other text (or on vacuum). Opening a book, he unfolds the inter-page space like a fan, which is a spectacular action in itself.

The inter-page and round-the-book vacuum presents a cosmos of literature of its kind. Is it possible for literature to come out of its cosmos? Naturally, yes. It is possible to create there its own platform bases which would maintain their connection with the page, it is possible to simply break through the page and insert it in the context of the events on the next page, and finally, it is possible to create with a series of kinetic actions a kind of living-space either between or above the pages (by applying stickers, by turning the pages, or placing big foreign objects on double-page spreads (like Meyer)

Beside communication through action, there is also a purely utilitarian approach to the book as an information carrier. The information carried by the tactile-thermal books, either heated or deep frozen, may be compared with the information we receive from surrounding nature which communicates with us through the weather.

Here we arrive at a conclusion that the aspect of book "reading" is not exhaustive in what concerns its capabilities. We come to a logical end: book can be everything. Any object may be used as a platform for a text, from a synchrophasotron to a TV set and a home refrigerator becomes automatically the design of a "deep" platform, to become an even more valuable contribution to extending the boundaries of fiction. By extending its limits almost to infinity, at the expense of integration processes, like supplements and substitutions, literature (through relations with theatre and art, music and mathematics and biology, as well as any other field of human activity), enriches the palette of its own expressive means, which had been brought to a logical end in the unfading masterpiece by Malevitch.

The "Black Square" by Malevitch is a rather stubborn and obstinate form of art. Poetry now comes to realise its personal role as a superpoetic monolith. Meanwhile any structure in art is found to be of an ambivalent nature, and, probably, is a continuation of everything else, including the obvious perfection reached by Malevitch. If we come to consider the "Black Square" masterpiece as a collapse of literature, which takes in all the words in all languages of all times and nations, re-entry to the verbal sphere becomes totally unnecessary. But then, probably due to that, other ways of poetic expression, like vector and gesture, come in view. The energy confined in Malevitch's collapse, as if in jail, asks for freedom of direction, that is for vector direction.

Futuristic letters put in motion, letters as such, have at least found implementation in gesture , in the movements of hands of the author himself. The implied interletter links, formerly called pauses, have become overgrown with invisible nerves of the visible vector of force.

Visual and phonetic literatures approach differently the pause as such. Formerly, the presence of pauses depended on the elements of a text. But of late, a tendency to inverse dependency is becoming more pronounced on text on pauses, their quantity and quality. In music the notion of a pause is developed better. (Chicherin has succeeded in formulating these problems in literature, but failed to work out a harmonious system of notions for vacuum.)

Vacuum, which is understood to be the absence of text, appears at the moment when the author resorts to the method of replacement of written text with such objects as pictorial art, or musical and even rhythmical content. "Self-induced" pulsation acquires universal value, which the literary rhythm has never had before. (For example, the verse by K.Morgenstern "Nocturnal Song of Fish", and many poems by Gappmayer.) By virtue of this universal nature of rhythm, a category common to many processes both in life and art, it is in the rhythmical field that integration of arts in one common rhythmical flow goes most successively. A verse composed of rhythmical proto-calligraphic elements may be called by right, both a musical, and an art, composition, and even a scientific outline.

Literature coming in contact with music also gets in touch with it on a common platform of breath (very characteristic in this respect is the cassette by Seiei Nishimura, which is a recording of his breath), and literature knows many such lines of interface with the various fields of human activity. (Although not every branch of science or art reciprocates such movement in literature).

Today, the world is filled with sonorous (or purely phonetic) poetry, and varied visual literary methods, whose enumeration only would take several pages. Visual-literary works (vacuum ones included) are found in plenty, both in the small "fancies" and the underground and in prestigious serious magazines and catalogues, like "Offerta Speciale" (Italy), "Doc(k)s" (France), "NIL" (Germany), "Solande" (Austria) etc. Besides, many other non-specialized magazines such as musical, mail-artist's and fiction, regularly publish visual - literary works (the Soviet "Iskusstvo" magazine, No 10, 1989, carries such materials). Yet among the variety of visual-literary works the publications completely dedicated to vacuum literature are still hardly noticeable. In the books by modern conceptualist poets, like Pierre and Ilse Garnier, Heinz Gappmayer and others, the vacuum problems are not accentuated, but rather put in line with other similar problems.

In the Russian literature, Vasilisk Gnedov, in 1913 in his vacuum "Poema Kontsa" ("Poem of the End"), was the first to personify the platform — which was like electrifying literature by turning on the white light of blank spaces in the dark of its print. (A propos, Meyer literally "electrified" one of his object-books by installing an electric bulb inside.) Thus, probably, already at the dawn of the century, Gnedov came to understand that literature is most closely connected with its own absence.

But there may exist a vacuum whose principal characteristic is not just emptiness, but rather FREEDOM from literature. If we penetrate deeper in the depths of vacuum, we shall see that any text applied to a clean sheet of paper presents a process of revealing text capabilities already present in this paper; and that means that vacuum literature is the freedom of the reader to suppose a text. In a wider view, any clean page has the capacity to hold an innumerable number of invisible texts, while the process of selective perception is the process of such pages watering-down.

The vacuum in which literature floats is its necessary reserve of pause consistence and territorial freedom.

In my opinion, the competition between vacuum and text improves the quality of literature in general.

Stamping The Mail-Out

D. Jarvis

Who sent this envelope? Nice stamps. Hey! That one shows Clinton's cat! Beautifully done: perforations, $ sign, elaborate border, gum and correct size. Amusing comment? Fine Art? Ephemera? The sender will be a mail artist. S/he has designed and printed an 'artistamp.' It fooled the Post -- they've franked it.

Mail-art (MA) is about a half a century old. Ray Johnson' correspondence is often taken as the first MA; or Duchamp, Schwitters etc? Emergence of the artists' stamp (AS) form is credited to Fluxus artists George Maciunas (1931 - 1978), and Robert Watts. From the outset it was subversive: vicious, funny, sexual, political, ego-centred and stupid. In that outer world where rich folk collect real stamps they have a term for unofficial stickers, labels and local post receipts: 'Cinderellas' Few philatelists specialise in them. My envelopes AS won't appeal to Cindy-hunters. They'd need to know edition size, dates, provenance, distribution details and how well known the artist is.

When is an artist not an artist? If you magnify that 3-point print on some of the world's stamps you may just see that it's by an artist of repute: rarely a living one; rarely a woman. Famous painters add class. They are culture-weapons used by boastful Nations. But these MA's are jokers. Their AS work sublimely ignores trappings of state, often by setting up 'alternative' worlds. Anna Banana of Vancouver puts out *International Art Post:* Full colour, perforated, gummed stamps, some by her, but most by other MA's. These artists are often outside the art establishment in their own lands.

If you've never seen an AS, the best way is to get involved on the MA network. There are rare gallery shows throughout Europe and North America. One, scheduled for May 1995, curated by Belgian MA Guy Bleus even has official backing. Postal authorities have sponsored AS events, but MA's are sometimes anarchic. They trick the system with 'variants' (='forgeries' to the Post Office) High craft seldom comes into AS production. Paper could be handmade, but it limits its use in photocopiers. It could be gummed, but sheets left around tend to stick together, unless peel-backings are used. It could be perforated but this requires special machinery. Pinking shears are an alternative. Harry Fox uses an unthreaded sewing machine for stamps, some of which are on paper that he makes. However, earliest postage stamps had to be cut out of paper and pasted onto the envelope by the postmaster and they look fine, so 'perfs' and gum do not an artistamp make!

Business letters franked in-house have their counterpart in the AS rubber stamp: either professionally made from the MA's Artwork or an eraser that s/he has hand-carved. Such images spill over many MA envelopes complementing or commenting on stamps, playing with your address and brightening the postie's day. This value-addedness isn't so easy with etchings, but you could thump the envelopes with a wood or lino block. Rubber stamps aren't just easier to use; they carry bureaucratic baggage --Kafka-- censorship, income tax, ...MA ridicules the State. Stamps question its self image: instead of monarchs or the Great and the Good, an AS deifies a mass murderer, saying: my country has also produced this monster, how about that!

Make a miniature drawing or painting, maybe give it a border, words, a number. Does this make it an AS? By its inclusitivity AS's answer is YES! and sending it through the post sanctifies it. But, such products are as ephemeral as the president's cat, and, by some definition, there is an awful lot of rubbish out there. MA shows prohibit jurying and entrance fees. No quality control? How disgraceful, complain High Art Critics. How shoddy, say Craftists; Ignore it!

Meanwhile the networking continues, letters, cards, artist's bookworks, zines, faxes, videos, tapes, writings, computer nets, 3D objects, and, prettiest of all? These artiststamps. At best they will have 'something to say' possibly serious, or they will make you laugh. Customs or Postal Officers may be less amused. As for collectors: they won't get to see them. And, in the spirit of Dada, Fluxus and a People's Artforum, perhaps that's all for the best.

Yet there have been moves to expose the As form to a wider public. Chuck Welch (aka The Crackerjack Kid) has organised several AS shows, the latest being 'Netshakers' 1994, NH, USA. With documentation that included a sheet of gum and perf stamps from some participants. CJK wants these stamps to be used, not to end up in forgotten libraries. Ed Varney, Dogfish, Joki, Harley, H-R Fricker and many others produce fine sheets and strange lands are created: Yamflug, Mraur, Tui-tui, Bao Badao, Ztzu, Tam, Aardon, Canadada, Doo-da, Terra Candella, Zaum, Edelweiss, Pan etc. There is a spirit of fun but also of outrage. The stamp responds instantly and publicly to blockades, tanker spillage, racial attacks, military coups, sex/gender persecution,famine, censorship, AIDS... Paintings and Murals take time and money to produce and expect great financial rewards. Capitalism.

MA taking something from Conceptual Art and Dada but much more from everyday life, condenses its ideas down to postage-stamp size and lets them float across the world exposed to all, seeking rapport not reward: The Artistamp.

Suggested Reading

lobach,O&B (1985: Weddel, Germany) Catalogue of Int. Artists' Postage Stamps Exhibition, MoMA Weddel. English Translation by Peter Kustermann.

Rastorfer, J-M (1980, 1985; Lausanne, Switzerland) Timbres d'Artistes I, II, (AS Artists Bookworks)

Varney, Ed. (1993) Foreword to 3rd Int. AS Biennial organised by JW Felter at Daidson Gallery, Seattle, WA, USA. (Unusual in that items were for sale)

***Welch, C.** (1994) As Int. MA Exhibit. AVA Gallery, Lebanon NH, USA.

Welch, C. (1995) "Eternal Network' MA Anthology. Univ. Calgary Press, ISBN 1-895176-27-1, $40 (A new survey, placing the As in its MA context.

TOP 107 HANDY ADDRESSES FOR BOOK ARTISTS

Compiled by John Bently

CONTENTS

BOOKSHOPS AND GALLERIES

1. St James' Art Books
Marcus Campbell
15 Piccadilly Arcade
London SW1
Tel 0171 495 6487

2. Dillons Art Bookshop
Andrew Lord
Long Acre
London WC1
Tel 0171 836 1359

3. Bookworks
Jane Rolo/Anna Pank
19 Holywell Row
London EC2 A 4JB
Tel 0171 407 1692

4. Eagle Gallery
Emma Hill
159 Farringdon Rd
London E.C.1
Tel 0171 837 1353

5. Hardware Gallery
Deidre Kelly
162 Archway Rd
London N6 5BB
Tel 0181 341 6415

6. I.C.A. Shop
12 Carlton House Terrace
London SW1 5AH
Tel 0171 930 0493

7. Workfortheeyetodo
Maggie Smith
41 Hanbury Street
London E1 5JP
Tel 0171 426 0579

8. Zwemmers, Whitechapel
80-82 Whitechapel
High St
London E1 7QX
Tel 0171 247 6924

9. Photographers Gallery Bookshop
Stephen Mears
5 Great Newport Street
London WC2H 7HV
0171 831 1772

10. Frith Street Gallery
Jane Hamlyn
60 Frith Street
London W1V 5TA

11. Printworks
Elly Robinson
45 Sir Isaacs Walk
Colchester
Tel 0206 562049

12. Museum of Installation
33 Great Sutton Street
London EC1
Tel 0171 253 0802

13. Off Centre Gallery
13 Cotswold Rd
Windmill Hill
Bedminster
Bristol
Tel 0272 661782

14. Camden Arts Centre
Arkwright Rd
London NW3
Tel 0171 435 2643

15. Matts Gallery
Robin Klassnik
10 Martello St
London Fields
London E8 3PE

16. Karsten Schubert
85 Charlotte St
London W1P 1LB
Tel 0171 631 0031

17. Counter productions
Ed Baxter/ Andy Hopton
PO Box 556
London SE5
(mail order, distributor of selected, inexpensive artist's books and much more besides.)

18. Greenwood Contemporary Books
Clive Warwick
51 Robertson St.
London SW8 3TX

SHOPS, GALLERIES AND DEALERS, U.S.A AND OVERSEAS

19. Art in Form
Janet Fryberger
2237 Second Ave
Seattle WA 98121
(shop selling A/Bs)

20. Art Metropole
788 King St. W
Toronto
Ontario MS5 1N6
Canada
(Distributes, collects, exhibits A/Bs)

21. Artworks
Barbara Pascal
11693 San Vicente Blvd
PO Box 158
CA 90049 USA
(Sells A/Bs)

22. Bromer Book Sellers
607 Boylston Street
Boston
MA 02116
(Letterpress books)
23. Buchhandlung
Hugo . M. Frick
Nauklerstrasse 7
7400 Tubingen
Germany
24. Califia Books
Edwina Leggett
20 Hawthorne Street
San Francisco
94105 USA
Tel 415 285 0314
25. Centre for Bookarts
626 Broadway
5th Floor
NY 10012 USA
Tel 212 460 9768
(publishes quarterly newsletter, send off for info)
26. Lyndsay Dobson Books
Box 285 Grimsby
Ontario L3M 4G5
Canada
27. Granary Books
568 Broadway
Suite 403
NY 10012 USA
(major seller of A/B)
28. Post Scriptum
Rue des Eperonniers 37
B-1000 Brussels
Belgium
29. Florence Locwy
Livres d'Artistes
46 Ave Rene Coty
75014 Paris
France
30. Martin Oleff
Posada Art Books
Rue de la Madeleine 27
Brussels 1000
Belgium
31. Musée Lilim
3 Rue Saint Jean
La Cité
11000 Carcassonne
France
Tel 684 72187
32. Volatile
Po Box 3274
Cincinatti
Ohio 45201
Tel 513 621 4976
33. Les Editions Belle Maleine
40 Rue des Saint Peres
75006 Paris, France
(Multiples by Artists)
34. Forum Book Art
Kompendium Zeitgenossischer Handpressendrucke
(Compendium of contemporary Hand Press Printing)
Heinx Stefan Bartkowiak, Ed
Kornerstrasse 24
D 2000 Hamburg 60
Germany
Tel 040 279 3674
35. Bookworks
407th St NW
Washington DC 20004
Tel 202 347 4590
36. Caroline Corre
14 Rue Guenegaurd
75006 Paris
France
37. Priscilla Juvelis
150 Huntingdon Ave
Suite 5DL
Boston MA 02115
USA
(Dealer in A/Bs)
38. Minotaurus Boekwinkel
Librarie Buchhandlung
Postbus 16477
1001 Amsterdam
(Bookshop selling and producing A/Bs)
39. Galerie Pascale Lesnes
56 Rue de Verneuil
75007 Paris
Tel 33145 49 04 157
40. Galerie Isobel Bongard
4 Rue de Rivoli
75004 Paris
France
41. Toni Zwicker
15 Gramercy Park
New York
NY 10003 USA
(Dealer in one-of-a-kind A/Bs)
42. Johan Deumans
Po Box 599
6400 AN
Heerlen
Netherlands
Tel (31) 710778
43. Walter Koenig
Buchandlung Ehrenstrasse 4
5000 Koln 1
Germany
44. Harry Ruhe
Galerie A
Johan Verhulstraat 53
1071 Ms Amsterdam
Netherlands
(Artist Books Gallery)
45. Printed Matter Bookstore At Dia
77 Wooster St
New York NY 10012
USA
Tel 212 925 0325

BINDING SUPPLIES, BOOKBINDERS, MATERIALS ETC.

46. James Brockman
High Ridge, Ladder Hill
Wheatley Oxford
OX9 1HY
(bookbinders tools, equipment)
47. Faulkiners Fine Papers
76 Southampton Row
London WC1B 4AR
Tel 0171 831 1151
48. Estamp
Silvie Turner
204 St Albans Ave
London W4 5JU
Tel 0181 994 2379
(publisher of books on printmaking and Artists' books)

49. Shepherds Bookbinders
Rob Shepherd
76B Rochester Row
London SW1P 1JU
(Supplies and binding
marbled papers)

50. N. Lawrence & Sons
119 Clerkenwell RD
London EC1
Tel 071 242 3534
(printmaking equipment,
tools, paper etc.)

51. Nautilus Press
Jane Reese
77 Southern Row
London W10
(Bookbinding, supplies and paper
mill)

52. Michael Putman
151 Lavender Hill
London SW11 5QJ
Tel 0171 228 9087
(Art and printmaking
tools and supplies)

53. McCarthy and Sons
310-326 St James Road
London SE1
Tel 0171 237 1946
(Cardboard Boxes and Containers
in short runs- any design, made
to order)

54. Paperpoint
130 Long Acre
London WC2 9AL
Tel 0171 379 6850
(large range of Wiggins Teape
papers
and monthly exhibitions
sometimes of A/Bs)

55. Parchment Works
Willen Rd
Newport, Pagnell
Bucks MK16 ODB
(Real parchment supplier)

56. John Purcell
15 Romsey Rd
London SW9 OTR
Tel 0171 737 5199
(Artist paper supplies)

57. Millway Stationers Ltd
Chapel Hill
Stanstead
Essex CM24 8AP
Tel 0279 81200?
(Cheap recycled stationary

58. Paper Back
Unit 2
Bow Triangles Business Centre
Eleanor St
London E3
Tel 0181 980 2233
(recycled paper supplier)

59. Svecia Antiqua Ltd
Robert Latham
Dowding Way
Tunbridge Wells
Kent TN6 3UY
(Specialist Paper Supplier)

60. Heyden & Son
Spectrum House
Hillview Gardens
London NW4 2JQ
Tel 0181 203 5171
(Bookbinding equipment supplies)

61. Atlantis Art Supplies
146 Brick Lane
London E1 6RV

62. Adana Ltd
162 Grays Inn Rd
London WC1
Tel 0171 837 6353
(Supplier and Shop: Lead type
Adana Letterpress equipment)

63. Paintworks
93 Kingsland Rd
London E2
Tel 0171 729 7451
(General art materials)

64. Rollaco Ltd
5 Witton Drive
Spennymoor
Co Durham
Tel 0388 818004
(Etching Press, relief press etc.)

65. Archival Aids
PO Box 5
Spondon
Derby DE2 7BP

66. Mould Type Foundry
Dunkirk Lane
Leyland
Preston
PR5 344

67. KA Office Supplies
Unit 2 Long St
London E2
Tel 0171 739 8765
(Cheap reconditioned office
machinery)

68. J.Hewit & Sons
Kinauld Leather Works
Currie, Edinburgh
EH14 5RS
Tel 202 966 9411
(Binding equipment, supplies
tools leather)

MAGAZINES AND PUBLICATIONS

69. Small Press
W.R. Crisp, Ed.
Colonial Hill RFD 1
Mt Kisco
NY 10549 USA
(Review of small press publishing)

70. Umbrella
Judith A Hoffberg, Ed.
PO Box 40100
Pasadena
CA 91114 USA
(News and reviews of A/Bs)

71. JAB
The Journal of Artists' Books
Brad Freeman
Interplanetary Productions
324 Yale Avenue, New Haven
CT 06515. USA
(203) 387 6735

72. Artists Newsletter
Publications Supplement,
Stephanie Brown
PO Box 23
Sunderland SR4 6DG
(Reviews A/Bs. Also publish
handbooks, directories etc)

73. Printmaking Today
Rosemary Simmons, Ed.
14B Elsworthy Terrace
London NW3 3DR
(Lots of reviews of A/Bs)

74. Directory of Specialist Bookdealers
(listing 740 Bookshops)
Peter Marcan Publishing
31 Rowliff Rd
High Wycombe
Bucks HP12 3LD

75. The Bookshops of London
2 Heads Publishing
Business centre
12-50 Kingsgate Rd
Kingston, Surrey
Tel 0181 974 8802

76. Art Monthly
(Cathy Courtney)
Suite 17,
26 Charing Cross Rd
London WC2H

77. Art Review
20 Prescott Plane
London SW4
Tel 0171 978 1000
(extensive coverage of A/Bs in 94)

78. The Printer
Box 1402
Findlay Ohio
45840 USA
(Letterpress printed news- paper)

79. Scottish Book Collector
11a Forth St.
Edinburgh EH1 3LE

80. Ambit
Martin Bax
17 Priory gardens
London N6 5QY
(Reviews of contemporary poets)

81. Creative Review
Dominic Murphy, Ed.
St Giles House
49 Poland Street
London W1V 4AY

82. Crafts
Geraldine Rudge Ed.
44a Pentonville Rd
Islington
London
N1 9BY

83. Modern Review
John O'Reilly Ed.
6 Hopgood St.
London W12

84. The Big Issue
John Bird, Ed.
4 Albion Place
Galena Rd
London W6 OLT

85. Frieze
21 Denmark Street
London W12H 8NE

PRINTERS

86. Instant Print West One
12 Heddon St
London W1
Tel 0171 434 2813
"the cheapest instant litho printer in Britain"

87. The Printing Centre
Store Street
London WC1
Tel 071 436 1593
(Cheap printing and typesetting)

88. Block Studios
The Paragon Centre
Searls Rd
London SE1
Te; 0171 708 5796
(Printing, bookbinding workshops Tuition, private work for individu-als,DIY Hire etc.)

89. Intype
Jennie Clark & Chris Shields
Input Typesetting Ltd
Durnsford Rd
Wimbledon Rd
London SW19 8DR
(Low cost short-run book service)

90. Community Copy Art
Dalby Street
Kentish Town
Tel 0171 267 9421
(Applemac and Copier facilities . DIY Hire)

91. Anthony Rowe
Bumpers Farm
Bristol Rd
Chippenham
Wilts SN14 6QA
Tel 0249 659705
(short-run printers and binders very cheap)

92. Reality Studios
Typesetting and Book Design
(for the small press, no job too big or small. Poetry a speciality)
Tel 0171 639 7297

93. Neals Yard Desktop
Publishing Studio
'the launderette of DTP'
(DIY DTP)
2 Neals Yard
London WC2
Tel 0171 379 5113

94. Kadacourt Ltd
Unit 11
Faradays Rd
Rabans Lane
Aylesbury, Bucks HP19 3RY
Tel 0296 96192
(Binders, finishers, shrinkwrapping, re-covering, slipcases)

95. The Book Factory
35/37 Queensland Rd
London N7
Tel 0171 700 1000
(Short run printers)

96. The Typography Workshop
Unit 33
Clerkenwell Close
London EC1R OAJ
(Letterpress printing for small fine print editions etc.)

97. London Print Workshops
421 Harrow Road
London W10 4RD
(Printing facilities and gallery)

98. West One Printers / Karizma Kards
35 Gt Pulteney Street
London W1R 3DE
("the cheapest colour postcard printers in the world" (but for how much longer?)

99. The Union Place Resource Centre
122 Vassall Rd
London SW9
Tel 0171 735 6123
(Printing DTP etc.)

100. Northern Writers Advisory Service
77 Marford Crescent
Sale, Cheshire
M33 4DN
Tel 061 969 1573
(copy editing, proof reading, type-setting)

ORGANISATIONS

101. Arlis (Art Libraries Society)
c/o Sonia French
18 College Rd
Bromsgrove
Worc B60 7NE
Tel 0527 579 298
(Publish annual directories & quarterly newsheet, often containing A/B reviews)

102. Private Libraries Association
Ravelston
Southview Rd
Pinner
Middx

103. Association of little presses
30 Greenhill
London N.W.3 5UA
071 435 1889
"formed in 1966 to bring together and offer mutual aid to small presses of all types. Subscription is £10 annually, for a bi-monthly newsletter, three issues of Poetry and Little Press Information (PALPI) and current catalogue. Our 25th Year!"

104. International Society of Copier Artists
Louise Neaderland
759 President Street * 2H
Brooklyn, N.Y. 11215
USA.

105. Northwest Review
369 PlC
University of Oregon
Eugene
OR 974403 USA
(regular reviews of A/Bs, features: send for details George Gessert)

106. Mainz Mini Pressen-Archiv
Fischtor Platz 23
D55116 Mainz
Germany
(Bi-annual Book Fair newly established as an archive)

107. The Small Press Centre
John Nicholson
Middx University
All Saints Site
White Hart Lane
Tottenham N17 8HR
Tel 0181 362 6058
(Advice, bookfairs, exhibitions, from the ashes of the Small Press Group...
Quarterly Listings, newsletter...)

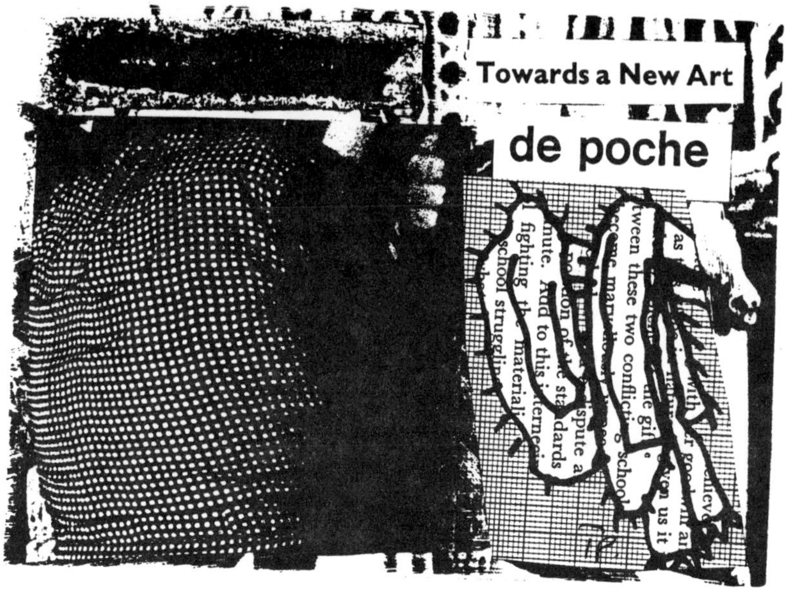

TI PARKS
TEN THOUSAND COLLAGES

WHAT SORT OF ANSWER IS NO?

John Nicholson.

If you want to be cured of worries about definitions visit Frankfurt International Book Fair. Here they have one simple aesthetic: money. A book sculpted in marble. Burned books as art objects. A book whose pages are six feet high flags fluttering in the square. Multiples aren't necessary with Art. The British have no time for Frankfurt. Just *after* it ends they announce the Booker Prize Winner. Is the Booker list too highbrow? Why should bookshops give them prominence when the books won't make money? Are shops public benefactors? How much does a window display or table-top cost for a promotion?

When has it been otherwise? I have never known it different. I publish but how can I sell when I can't relate to the 'book trade'?

Ten years ago I reconciled myself to the fact that no matter how 'popular' I made my publications, I wouldn't sell thousands. Why not? I could make them cheaper than the average mass paperback. I could make them better looking. The sales didn't vary. I could not promote them so their chances were limited inevitably. No window display or table-top. Yet I sometimes sold more copies than the Booker titles! But nobody took notice because I was not a member of the club. I belonged to the other club, the self-publisher who was below the Vanity Press. At least the latter are recognised as mugs.

But think positive. People like me are not backward in determination. Make a virtue of necessity. I would make limited editions. I would make a good-looking hardback. At four times the price. What happened? It not only sold well, it sold better! It won plaudits and cash.

I launched it with fanfares of a different sort. I organised a five-day event around it. Every day there were three talks by experts in their subject. Each day themed and accumulated. There was a permanent exhibition and a bookstall selling The Book and related titles. A tea party was the official launch by invitation only. They fought to buy signed copies. By the end of the week I had sold so many copies I had covered my print bill and the hire of the space. The momentum carried into reviews and made it a Book of the Year in *City Limits.* But it took nerve and imagination, I used every trick I could and every contact. It got me a distributor who sold less copies than I had. It congealed once it was dealt with properly.

But I saw what was possible. I could see that there were others like myself doing similar stunts. All of us held no illusions about a best-seller nor did we hold scruples. We all knew what we were doing but we didn't know each other's tricks. None of our subjects overlapped so we could attract a different audience, many audiences. Likewise with the media. We offered them not just one novel angle but dozens! I saw that if we joined forces we could make a huge splash!

Many of us already helped each other informally by including each other's bumf in our mailings. It was casual and friendly. None of us thought of becoming dealers which, in retrospect, was the next logical step. Perhaps we were too lazy or just couldn't care. That would have made it a business.

The catalyst was an epiphany in America. There we discovered people like us had organised. They came together annually — not just to have a beano of swapping and gossip but also to show and *sell* to the public. And because the event was successful and unique the media sniffed around. What was going on? Was it a phenomenon? How to describe it? (NOT define it.) Actually we hit the death throes although we only realised that later.

The American small press fairs were becoming victims of their own success. Because sales were worthwhile for some participants they were taking them seriously. Small publishers who already took themselves seriously swallowed their pride and exhibited too. As there was money being made scouts came to pick off the winners. Commerce was stalking and the nature of the event was changing as well as the views of the participants.

But we saw none of that. Instead we saw the original vision unspoiled. Therefore the British small presses benefited from our moment. We saw what was possible in both ways. We saw how it could succeed and fail. We saw a heady brew of the most amazing ingredients.We saw the inside aspect as well as the commercial 'sensible' side. We saw that both could live together as long as neither dominated. All that was needed was to extend our formal practices slightly. After all it is a rare small press which doesn't mix with like-minded people so every small press knows at least a couple more. And since the impulse was creative rather than commercial, small press people mingled with other creative people. 'Small Press', as we used the term, was unlimited. It drew in and out.

For the record our first meeting together was held in John Michell's flat where we told them about the American way. We knew we could put on some show ourselves if only the presses in the room co-operated. Who was there? John, Richard Adams, Bob Rickard, and Paul Sieveking of Fortean Times, Alistair Brotchie and Antony Melville of Atlas, Val Stevenson of Women in Publishing and Celia and myself.

On the face of it you couldn't arrange a more contrasting crew or a wider cross-section of interests. Yet in practice we were all on the same side. We all shared the same contempt for the book trade. We shared the same ambition: to do the best job possible in presenting our publications. This set us apart from the perception of little publishers, many of whom were poets! If we held a shared fair there would be one certainty: no poetry readings! There might be all sorts of strange goings on which would be different from predicatble book launches but NO POETRY RECITALS.

In September 1987 we met for the first time and decided to work together in an organised way. We would be an organisation, a group. The name was easy and the acronym had resonances: SPG? Special Patrol Group, Society for Propagation of the Gospel? We would be the Small Press Group.

It could not fail to be exciting. It would be an adventure. We respected each other and actually enjoyed each other's publications. It was fertile. It was not provoked by business considerations. It was entirely business. That was September 1987. A year later the Small Press Group of Britain had held its first annual fair and launched its first Yearbook.

ARTISTS' BOOK FAIRS IN FRANCE

The artists' book is a well established genre in France with a history of both small press publishing and editions made by large galleries and dealers. It is not necessary, as it is in Britain, to explain what an artists' book is to most people in the art world and many small public libraries have some representation of these books, if only by local artists. It is interesting to note that there is even a public lending library of artists' books at the Bibliothèque Municipale in Roubaix and a large collection of small press publications at the Bibliothèque Municipale in Marne La Vallée near Paris, useful if you are on a trip to EuroDisney! Most major cities have a central library which includes often excellent collections of artists' books, amongst the best are those at Lyon, Toulouse and Bordeaux. There is of course the almost unsurpassable collection at the Bibliothèque Nationale de France in Paris which includes, due to the national depot, nearly every major artist's book published in France as well as an extensive collection of foreign publications.

Artists' books are often exhibited at the numerous *artothèques* established throughout France as well as more specialised venues such as the Maison du Livre d'Artiste Contemporain in Domart-en-Ponthieu near Amiens in Picardy. The Musée de l'Arsenale de Gravelines often has exhibitions of artists' books as well as graphics and is only a few minutes drive from Calais.

There are a number of specialist book fairs covering the general area of artists' books at various times of the year and spread throughout the country. One of the most established is the Biennale organised by Pays-Paysage formerly of Uzerche in the Corrèze and now at St-Yrieix near Limoges from 2 to 5 June 1995, this fair includes exhibitions and discussions and a training day for librarians and curators. Pays-Paysage are currently initiating Le Centre du Livre d'Artiste also at St-Yrieix. L'Art à la Page is the name of the fair at Cagnes-sur-Mer on the Mediterranean coast near Cannes, one of the older fairs which in 1995 runs from 3 to 5 June, an unfortunate clash with St-Yrieix. There is a selection process for L'Art à la Page but the stands are free, and exhibitors are responsible for their own board and lodging. Le Marché de la Poésie takes place every summer in the open air in the Place St Sulpice in Paris and, by tradition, it never rains. This fair runs from 22 to 25 June 1995 and mainly attracts publishers of multiple artists' books and small press poetry, this includes everything from small limited editions to xeroxed sheets, it does not generally include *livres d'artistes.*

There are two small fairs organised by Editions Tarabuste, both in the west of France, the first is at St Benoit du Sault and is the Rencontre des Poèts et des Artistes which normally takes place in early June, it is only open to invited exhibitors and includes readings and other events along an annual theme; 1993 featured Les Petites Editions du Grand Pirate and included readings by Serge Pèy; 1994 saw a tribute to the Danish sculpter Jacobsen. It is well worth a visit with its casual party atmosphere. The second is a new fair L'Espace du Livre which is to take place 19 to 21 May 1995 at La Roche sur Yonne and will be a larger fair including workshops and demonstrations.

There is also another form of book fair popular in France, where you send your books for display at a small charge. These fairs include the Soleil sur la Page which is organised by the Edelweiss Noir Association at the Espace Liberté in Crest 31 March - 17 April 1995 on the theme of La Littérature et Le Cinéma, and the Salon de la Petite Edition which takes place from 2 to 5 February 1995 at Quimper in Brittany.

The most adventurous project of 1994 was the 1st Artist Book International which took place at the Intercontinental Hotel, Paris from 1 to 4 December. This fair includes (at the time of going to press and if the organisers are to be believed) some of the major names in the artists' book field, however, there seem to be some glaring omissions due to an eccentric process of selection/invitation. It was also rather expensive to participate and therefore appeals more to the luxury end of the market. It is planned that it will take place alternate years in Paris and New York, it is certainly exciting to see a major initiative, like this one, in Paris. It is also possible to see and exhibit artists' books at the SAGA, also in Paris, which has currently de-camped to the Parc d'Expositions at Porte de Versailles from the crumbling and dangerous splendour of the Grand Palais, which is closed for major repairs. This is also a very expensive fair and prospective exhibitors are subjected to selection. As it is a fair for all editioned work it is interesting to see the artists' books in the milieu of the art world in general. The future of the SAGA is however under threat. It is possible that in 1995 it will be amalgamated with the FIAC art fair and publishers will be "invited". This could mean that it will go the same, unfortunate, way as the excellent "Edition" that ran for three years concurrently with Art Basel in Switzerland before being merged with the main fair after pressure from a number of influential powers in the art world who seem to have felt threatened by the success of the cheaper editioned work. This reduces further the limited number of venues for the display and selling of editioned work to a larger audience.

It has to be said that by far the best way to sell one's publications is still to go around all the various collections personally and there is always a sense that many of the fairs above are really only for exhibition. Some cost little or nothing to participate in and others are so expensive that even the most wealthy publishers have to think long and hard before taking part. These fairs nearly always offer an opportunity to meet other publishers and artists and one regularly finds new clients and discovers valuable pieces of information one didn't have before. For further information the following addresses have been included for the purposes of direct enquiries.

Matthew Tyson and Isabella Oulton of Imprints.

Adresses for enquiries to artists' book fairs in France:

Pays-Paysage
15 rue d'Alger
87000 Limoges

L'Art de la Page
Bibliothèque Municiplae
Avenue Maurice Donat
06800 Cagnes sur Mer

Le Marché de la Poésie
Agence AD'HOC
6 rue Notre Dame de Nazareth
75003 Paris

L'Espace du Livre and
Rencontres Artistes et Poèts
Editions Tarabuste
rue du Fort
36170 St Benoit du Sault

Le Soleil sur La Page
Association Edelweiss Noir
Espace Liberté
5 rue des Alpes
26400 Crest

Salon de la Petite Edition
Maison pour Tous d'Ergué-Armel
4 rue Saint-Alor
29000 Quimper

1st Artist Book International
46 rue de Sévigné
75003 Paris

NO JACKET REQUIRED

ROY MARCHANT

As a producer of artists' books, I would unquestionably give my full support to actions that would bring our work to the attention of a wider audience. However, I feel that the growing popularity of artists' books has to some extent made them a victim of their own success.

This became clearly evident at the 1994 bookfair, at the Royal Festival Hall. Although obvious improvements had taken place over the previous year, it would seem that the inherent values that separate artists' books, as a discipline have been conveniently forgotten.

I can openly confess, as much as anybody, a certain degree of "blind love" for the subject, but what message then, if any, does this convey to an often dismissive public? Can artists' books ever be more than well-crafted objects to adorn coffee tables or a box of prints one can selectively frame? Although my answer to this would be a defiant 'Yes', I could sadly only find this applicable, in my opinion, to a small percentage of work.

It would appear that in this race for popularity, the identity of artists' books has lost its way or at least taken a wrong turning. The founding ideas of innovation, of "hot art" have been exchanged in preference for a pair of white gloves and a selection of handmade papers.[1] We cannot try hard enough it seems to make each work more inaccessible than the last. The notion of publication has been eclipsed by that of a mere production for production's sake. I once attended a lecture at the Slade School whereupon I was told that we were in danger of "Making Art and Killing Culture", eight years on this would be taking matters to an extreme, but I feel intentions have to be made clearer.[2]

An artist's book must go far beyond being subservient to painting or printmaking. Being nostalgic over such a subject may seem somewhat of a contradiction, but the pioneering work of artists such as Edward Ruscha, in the early 1960s is too important to be lost in the convenience of art history.

26 Gasoline Stations and subsequent publications may in today's climate appear dated but as a body of work I feel that their contribution is unequalled. Ruscha echoed America's shift from literary culture to car culture and in the process created one of the world's first artist's books

So, does any discipline progress past a certain point and then merely produce more of ? Possibly, but with the ever promised extinction of the printed page surely this point can be moved from that of the mere cosmetic. This only creates a false sense of well-being.

I feel that the artist's book has a very positive future in terms of the contribution it can make to contemporary art. This is only possible if it is allowed to evolve into a fully independent discipline and not balance itself around the periphery. If this discrimination is not made we shall continue to applaud ourselves for merely selecting the correct paper.

1."hot art" conversation with Ed Ruscha, Bernard Blistene 1990

2."Making Art Killing Culture" lecture, by Slade Prof. Bernard Cohen, 1986

Collective Copy:

Fritz Eipok

Imagine a regular magazine without an editorial team or advertisements. Collective Copy is just such a project, recycling information in its purest sense by making unique use of a mass of information and linguistic material.

Jurgen O. Olbrich, a collaborative artist, founded Collective Copy in 1976 and it has remained unchanged since. He simply puts waste-paper baskets next to photocopiers in public places, collects all rejected copies, stores them for a year and then divides then into two categories.

One category, the 'Archive of Lost Information', contains indecipherable copies which one day, he hopes, will be decoded with the invention of a machine to extract hidden information from them. Most of the copies, however, contain some fragments of visual information, which Olbrich makes use of.

Collective Copy is published quarterly, and the concept behind it is explained at the front of each copy. Subscriber's names and addresses are also printed so that they can contact one another. Each copy is different, and each subscriber will receive only one. So, if you would like to receive an original art magazine, contact Jurgen Olbrich at the address below and be prepared to get a surprise in the post.

Jurgen O. Olbrich, Bodelschwingstra. 17,
D-34119, Kassel, Germany.

BOOK ART IN SCOTTISH COLLEGES:

INFLUENCES AND PRACTICES

Marshall Anderson.

During the 70s and 80s the Henderson Gallery and the Graeme Murray Gallery provided a showcase for artists' books in Edinburgh, establishing interest in this vehicle for expression and accessibility. A Papermaking Symposium at the Fruitmarket Gallery in November 1980 provided an historical link between 70s America and Edinburgh, triggering the teaching of book-making at Edinburgh College of Art. George Donald, inspired by the symposium and a subsequent visit to Twinrocker (Indiana Papermakers), established a paper making facility in 1984 within the printmaking department at ECA and invited Tom Simpson, a trade binder of Jordanstone College of Art, Dundee, to demonstrate binding techniques. These were based on traditional stitch bindings, thereby setting a precedent for the way book making would be taught through that department in the future.

After a gap of ten years, Faith Shannon, a fine art binder working in Argyll, was invited to ECA's printmaking department in February 1984 to give a workshop. Again very traditional skills were shown to students. For Elspeth Lamb, who had initiated the invitation, it is important for students to get a solid base on which to build. "Without that," she says, "it's harder for them to get to grips with conceptual ideas." As a consequence of this, students in, and from, the printmaking department of ECA tend to produce books which fold, hinge and bind together printed images and text. In the tapestry department a different story is evidenced.

Tapestry weaver, Gordon Brennan, Started teaching at ECA in 1986 where he introduced book making to a few students. He had been inspired by visits to the above mentioned galleries as well as Ulyses Carrion's *Other Books and So* in Amsterdam. Brennan was not as concerned about traditional binding techniques as his peers, instead books were about collecting and collating ideas. Book-making is now introduced to second year students as soon as they begin to specialise in tapestry and they are encouraged to go their own way, extending the practice of gathering data and rationalising binding for themselves. Some then continue to assemble ideas in this way through third and fourth year. Brennan also teaches in the Department of Constructed Textiles at DoJ where his philosophy provides an exciting counterpoint to more traditional approaches favoured in other departments. In 1986 Jonathon Robertson took over photography where he introduced basic book binding. "With simple tools and materials you can produce a cultural object", states Robertson, who encourages students to treat the book as an intimate medium. This reflects directly upon a set project documenting self and family and pursuing ideas relating to personal identity. Most students at DoJ have access to photography, so anyone can attend Robertson's basic binding workshop. However, he prompts students to use the bindery at St Andrews University, now affiliated to DoJ, which provides a cheap service for all students and staff.

When Thomas Joshua Cooper came to Scotland and established the Department of Fine Art Photography in Glasgow School of Art in 1982, he brought with him contemporary American art history and artists' books which were placed firmly on the agenda. Graeme Murray, who had published books in collaboration with artists since 1976, was called in as a visiting lecturer bringing to the course notions of the book as part of an aesthetic currency. An item to exchange with, a form of presenting an artist abroad. Students were encouraged to experiment: one made a book in a suitcase, some did walk-in-books, there was a kite book, and one made of glass. Simon Cutts and Erica Van Horn also visited and Tom Valentine, a commercial binder and book restorer, was invited to give workshops employing traditional techniques and other (in-house) lecturers included papermaker, Jacki Parry.

Lecturer, Vaughan Judge, explained that students are introduced to the concept of the book as an extension of the gallery space in their second year. The book, or catalogue, is seen as a supplement to both the gallery and their exhibition. He admitted that to begin with the book was seen as very limited, unique, or precious object, whereas now the students are versed in more commercially accessible printing techniques such as xerox. David Bellingham joined Fine Art Photography staff (after being a student there and influenced by Graeme Murray) in 1990 bringing a pamphleteering philosophy inspired by the bookworks of Ian Hamilton Finlay. Bellingham sees the book as a ready form for distributing and exchanging ideas through established art networks. For him the book is not an object handled with cotton gloves but an immediate carrier of ideas — mass produced — disposable.

American born and trained, Mary Modeen, began lecturing at DoJ in 1989 where she was part responsible for setting up Fine Art Printmaking. "The 70s bookmaking thing in the US touched everyone who was working in college departments," she said. And making books provided her with a complementary and egalitarian way of presenting lithographic images and text. "It's something that's a container of ideas." — a maxim that's imparted to her students. Tom Valentine provides an introduction to basic construction in folio and lipped clamshell boxes but not stitch binding. In third and fourth year the course is student led. Modeen and her associates are there to provide knowledge, to bring in appropriate instructors, to send students away for extra courses if necessary. Her students produce an exciting eclectic barrage of work.

Scottish artists who have made books since the 70s and who acquired no technical knowledge of the practice in art school evolved, through their own logic, book designs and associated necessities. Some have extended the practice into installations, others devoted time to precious one-offs, a few employed the form of large editioned, printed multiples. All discovered the same problem, and it is one that has never been solved. There is still no properly organised facility for showing artists' books in Scotland. More students in Scottish colleges are showing an interest in book-making each year, but the crucial lack of an exhibiting space often deters them from pursuing book practice. If Scottish art schools continue to open students' minds to the potential of bookworks then these students, who are on the verge of taking the medium towards new frontiers, are going to continue to feel let down by conservative gallery administrators, who fail to provide the much needed forum for debate and development.

AND
THAT
THERE...

Simon Redington

ARTISTS' BOOKS AND PUBLICATION ACTIVITY

AT THE

UNIVERSITY OF DERBY SCHOOL OF ART AND DESIGN

Research Group for Artists' Publication

A research centre for artists' publications has been established at the University of Derby, and is based in the School of Art and Design. Known as RGAP (Research Group for Artists' Publications), this group comprises members of staff from Fine Art, Photographic Studies, and Graphic Design. The aims of this research group could be summarised as follows:

To establish a centre at Derby which will promote and publish artists' books/publications/editions/multiples. This involves both artists working at the University, and invited artists.

To build up an archive of contemporary artists' publications for reference and research, (housed in the School of Art and Design library).

To promote discussion of the concept of artists' publications by means of conferences, seminars, practical workshops, demonstrations, and exhibition.

To collaborate with other centres of such activity in the UK and abroad, with a view to establishing joint projects, publications and exhibitions

To offer assistance and expertise in the production of publications.

Funding is available for the above activities via University Funding mechanisms.

One of the main achievements in 1994 has been the selection by the European Commission of a project titled *'European Artists' Books / Publications'* for a new initiative in arts development and education in Europe. This project was written and submitted by Martin Rogers, Director of the Research Group, and has subsequently been supported and co-ordinated by other members of the group. The project is primarily concerned with developing a European dimension to the activity of making artists' publications, both for students and professional artists. The research group at Derby has been leading this project, and will continue to co-ordinate activity beyond the first phase of the project.

The structure of the project provides a vehicle for individual / collaborative publications to be made, with distribution to a network of European libraries in art schools, archives, and collections. Although the project is Europe-based, some archives and collections in the USA will also receive the published material.

The project commenced in June 94, with meetings of the international partners at the University of Derby. The other partners currently involved are Crawford College of Art and Design, Cork, Eire; Academie voor Beeldende, Kunsten, St Joost, Breda, Netherlands: and Chelsea College of Art and Design, London.

Funding for this project has been provided by the European Commission.

BA Hons Fine Art ***'Editions and Publication'***

A new area of study has been established on the BA Fine Art Course at Derby, and is known as *'Editions and Publication'*. In their 3rd year, students can opt to work entirely and specifically in this area. Work may take the form of books, publications, multiples, objects - in fact in any form that is capable of replication. The book is seen as a viable alternative space for the dissemination or 'exhibition' of ideas. Students propose and discuss appropriate forms, formats, and production methods for their work, and are expected to define a clear critical attitude to the making of work in this idiom.

Contact: **Martin Rogers**
School of Art & Design, University of Derby, Britannia Mill, Mackworth Road, Derby DE22 3BL
Tel: 0332 622222

1983 This is an important major investigation for me — these three books — but no-one has attempted to review this project.

PRAXIS

my signature

Collaborations, Stefan Szczelkun, 1987. Asks if working class culture can produce serious art, with visual reports and original documents from the artist's collaborations. Price: £4.50.

← challenging the stereotype that w/c art is shallow + simplistic

work as editor!

Some people might say they're not artists books

paperbacks illam

Class Myths & Culture, Stefan Szczelkun, 1990, ISBN 1 870736 03 5. Collection of essays addressing issues of class and culture, with reports on three large-scale collaborative performance art events relating to urban communities. Price: £5.

This ad was for my ... I'd have to keep my books in the ...

unique
A ~~theory~~ of 'Glamour' in here which I want to develop.

the three books all connect up — Also very dense

I publish because these things are not being said or got away ... — not really

would like to have extended this with addition of Lord Keith (BBC)

The Conspiracy of Good Taste. Stefan Szczelkun, 1993. William Morris, Cecil Sharp and Clough William Ellis are scrutinised to begin to understand the mechanics of cultural oppression in various media. Includes an historical tracing of the history of good taste and a searing critique of Bordieu's sociological exposition. Price: £9.95. hardback.

Cheques made out to WORKING PRESS →
P + P free in ~~....~~. U.K.

CENTRAL BOOKS

note: new WORKING Press address! 54 Sharsted St. SE17 3TN

THEORY

Thanks
1993 Tanya

TONY KEMPLEN
29 Meadowhead
Sheffield S8 7UA
Tel 1742 746091

My artist's book **SK329858** takes the form of a pack of playing cards. It recounts the wartime crash of an American bomber (nicknamed the Mi Amigo) in a Sheffield park in which ten airmen were killed. The images used, relate both to the historical facts and to the site as it is now, 50 years later. The presentation as a deck of cards allows the often random nature of war and fate to be explored. The cards may be arranged into groups of nine cards to make six further images. The pack may be seen as an alternative memorial to the dead Americans as well as an exploration of the effect of the crash on the collective perception of the site fifty years later. The pack comes in a box with a 16 page booklet containing an introduction by Sharon Kivland and explanations of the history and images used.

Edition of 100. Price £10

SOPHIE ARTEMIS
Moonacre House
Cardyke, Blankney Wood,
Metheringham,
Lincoln LN4 3AN

Identity motivates my work: the identity of people, places, words, objects and moments.

In my books the reader is involved in constructing the idea contained in the book. There is movement, texture, light and weight in each piece, and through the layers, like a palimpset, an individual moment or point emerges from the other contextual information

I make unique, one-off books.

BRIGID McLEER
32 Crescent Grove
Clapham Common
London SW4 7AH

The books are small, intimate musings on the relationship between experience and its translation, which is the creative impulse. They use the form of the artist's book, a collaboration of word, image and object, to explore this relationship. They aim to give meaning but also to question the nature of this process.

translationships
1994. 8 x 5 cms approx. Book of images on acetate and paper. £120.

experience...and the view
1994. Four paneled, double sided fold-up book. 16 1/2 x 11 cms x 4 panels. £135.

SENSE (newspaper)
1994. 30 x 12 cms.
Images on ruled recycled paper. £35.

Untitled
1994. Green box containing photographs and books. 30 x 16 x 3 cms approx (box). £275.

The book as art

This is a new piece of book art constructed by Les Bicknell. Its form and content question and ponder the role of the book, its history and its future. Five specially commissioned essays from leading voices in the book art field explore the notion of the book as an object. Essential for an understanding of the medium.

Edition of 150, signed and numbered copies
330 x 320, opens out to 970 x 320
8 colour litho on recycled paper within grey board covers.
£25.00

Order from:
the book as art
Eva's Place, Sibton Green,
Saxmundham, Suffolk IP17 2JX
0728 668639

LIST OF ARTISTS' BOOKS PRODUCED BETWEEN SPRING '94 TO '95

(Max of 3 books listed per artist)

Aard Press,
c/o Aardverx,
31 Mountearl Gardens,
London SW16 2NL
Auntiesdisoestrogenistablishwimmenta rhythmaniacism
Dawn Redwood
Images, words, of blocked-ear giddy misery; hissed continuous panics. Printed on sickopieragain with cheapie paper, stamped, labelled & unstably stapled. First version ripe for mutilating and later hating as a reminder of bad times. Why not antidisestablishmentarianism?
1st April 1995, London
Height: 10.5, Width: 7.5cms. Edition of 20. £5.
πr2 and Pie-eyed
Dawn Redwood
Semi-circular bookwork. Bits of mathematics and unrelated imagery collage. Red and black on coloured papers.
31st April 1995, London
10.5 x 21 cms (diameter). Edition of 50. £4.
Uro-Doll
Petal Jeffery
Livre d'Artiste. 7.5 hand-coloured paper-dolls, each unique, based on the artist's silhouette. Fragment of "tea dance outfits list" on the reverse. No other text. Purplish cover.
21st April 1994, London
12.6 x 12.8 cms. Edition of 12. £12.

Sophie Artemis
Moonacre House, Cardyke,
Blankney Wood,
Metheringham,
Lincoln LN4 3AN
Hidden Possibilities
Explores the idea of exposing and hiding situations and moments: for instance, there are two handprints, placed on opposite pages so that they can only ever touch, unseen between the pages or when the book is shut.
1994, London
29.5 x 20 cms. Edition of 2. £100.
Remember
A glass box, etched with words relating to memory and communication of memories. Inside this there are eight different loose 'pages' made from discarded objects and materials, each bearing a letter from the title.
1994, London
21 x 15 x 8 cms. One-off. £250.
Icarus
In a feather and wax covered box lined with gold is a feather-covered book. Within the wooden, wax and feather covers are pages, ultramarine one side, gold the other, folded like a concertina. When the pages are opened out each has a letter cut from it: "I-C-A-R-U-S"
1995, London
8 x 16 x 6 cms. One-off. £150.

David Barton
45, Wellmeadow Road,
Hither Green, London SE13 6SY
50 Strips
Linked sequences of line drawings about the human figure, laid out in strips across the page; each segment representing a glimpse of the body's struggle to occupy/vacate the space which contains it.
July 1994, London
A4, 30 pages. Edition of 50.
ISBN 0 907559 65 4 . £6.
Stunt 2
The second in a projected series of five books containing sequences of line drawings in which the human figure strives to extend/transform the limits of its own confinement.
November 1994, London
A5, 32 pages. Edition of 500
ISBN 0 907559 68 9. £4-50.
Stamping
Sequences of line drawings, searching for the continually changing image which "I must find and grasp again and again and again, day after day if I am to continue to become the thing I am."
January 1995, London
A5, 336 pages. Edition of 500
ISBN 0 907559 62 X. £15.

Book Works
19 Holywell Row,
London
EC2A 4JB
Rex Reason
Simon Patterson
The form of the book is based on the chemical periodic table. This established form is significantly altered by the insertion of famous names. The book should be seen in close relation to Patterson's other work.
13.5 x 10.5 cms. Edition of 2000
ISBN 1870699 13 0 £6.
Two Oxford Reading Rooms
Joseph Kosuth
The book contains two texts 'The (Ethical) Space of Cabinets 7&8' and 'Say I do not know' which originally formed part of two installations at the Taylor Institution and Bodleian Library, Oxford University. Commissioned by Book Works as part of The Reading Room
9.8 x 21 cms. Edition 1000 standard, 250 hardback, signed and numbered.
ISBN 1 870699 14 9. £10 (standard) £75 (hardback).
J'appelle un chat un chat
Sharon Kivland
A response to an extract from Sigmund Freud's Case Histories 1 'Dora' The texts in the book are the

names of lipsticks, each a shade of red, but may equally be read, without prior knowledge, as referring to states of emotion or feminine attributes. The book is hand printed and bound in black calf leather.
21 x 21 cms Edition of 10, signed and numbered.
ISBN 1 870699 16 5. £550.
A video by Shelagh Wakeley (ISBN 1 870699 16 5) has been produced in response to this book and the first ten copies are signed and numbered and presented with the bookwork.

Frans Baake
Auskamplanden 2, Enschede, Netherlands
Couleur Locale
6 Time-saving photo albums in a wooden box. Although just 7mm of each photograph can be shown (the pictures have been xeroxed in colour), it seems to be clear that each remote corner that the artist has visited has its own atmosphere.
1994, Enschede. Netherlands.
26.5 x 17 cms. Edition of 7. £375.
Assez Enough
A square book, containing black and white three ply prints in combination with (hardly visible) silk-screen prints. The question is still unanswered: Are the contents of the book enough by means of quantity, or assez in quality?
1994, Enschede.
29.7 x 17 cms. Edition of 8. £90.
Same Spot with me
A leporello in a linen box, containing some pictures of f.1. the Moscow Red Square, as published in an authentic travel-guide, in combination with a snap-shot of the same location with the artist on it. Will it ever be the same after the artist's visit?
1994, Enschede
20 x 33.5 cms. Edition of 5. £250.

Victoria Bean
119 Lansdowne Road,
London W11 2LF
(E) Motions of Silence
A sculptural book in a box, letterpress.
26 x 18 cms. Edition of 24. £60.
The Foundling
A Sculptural book, letterpress, varnish and masking tape.
27.5 x 19 cms. Edition of 10. £60.
A Place to Hide
A sculptural book in three pieces, letterpress. Slipcase and lead weights.
October 1994, Circle Press.
29.7 x 18 cms. Edition of 14. £90.

Les Bicknell
Eva's Place,
Sibton green, Saxmundham,
Suffolk IP17 2JX
invisible words
The stuff you construct while waiting for people, plus the stuff you overhear on the bus, together via the photocopier. Within an experimental binding , with instructions on how to fold it.
Various sizes, Edition unlimited. 10p plus p&p at time of ordering.
mapping knowledge
An experimental bookwork with challenging typography to accompany the exhibition of the same name. Containing artists' responses to the question "why do you make books and why are books important to you?"
1994, the Minories gallery, Colchester.
1/2 A4 when closed, when open variable. Edition of 150. £15 plus p&p.
the book as art
Les Bicknell
Richard Minsky, Cathy Courtney, Matthew Tyson,
Ivor Robinson, Keith Smith.
A piece of book art. Its form and content question and ponder the role of the book, its history and its future. Specially commissioned essays from leading voices in the book art field explore the notion of the book. Essential for an understanding of the medium.
1994, Essex University.
33 x 32 opens out to 97 x 32 cms. Edition of 150. £25 plus p&p.

Roy Bizley
12, St Johns Road,
Stoneygate, Leicester, LE22 2BL.
Seasonal
A smaller format section at centre of book, carrying a 12-line poem with monotypes. The larger end papers, supporting images, heavy rigid paper cover, all printed from wood and lino, black plus colours. Hand sewn, 1994. 16pp.
21 x 30 cms (open) Edition of 10. £6-75. incl. p&p.

Burning Book Press
182 Worple Road,
Wimbledon, London SW20 8PR
Untitled (volume 1)
Danny Flynn
Handprinted letterpress, containing word deletions from manuscripts of T.S. Eliot, François Mauriac, Ernest Hemingway, Allen Ginsberg, Truman Capote, Robert Graves, William S. Burroughs. (Hardcover) 10 pp
February 1995.
6.5 x 13 cms. Edition of 100. £12.
Alphabet Soup
Ian McAlister
Handprinted letterpress, originally designed as a mailer for a lettering company. Contains 3 pictures printed in salt, pepper and tomato soup, with enamel badge.
January 1994, London.
9.5 x 7.5 cms. edition of 100. £6-50
Public Dreams. (The work of Jeff Higley)
Jeff Higley
13 loose leaf single colour handprinted letterpress photographic descriptions of the Green Man, Tree Dressings, Installation Work, Sculpture. (Soft cover) 14pp.
January 1994, London.
19.5 x 31 cms. edition of 500. £12.

THE EAGLE GALLERY

159 Farringdon Road,
London EC1
Curated by Emma Hill
0171 833 2674
Gallery Opening hours are
Thurs/Fri 11-6pm, Sat/Sun 11- 4pm
or by appointment

The Eagle Gallery was established in September 1991 to exhibit and promote the work of British artists through exhibitions and a regular publishing programme of limited edition artists' books and portfolios. The Gallery stocks a large selection of bookworks by artists and other private press publishers including Les Bicknell, Natalie d'Arbeloff, Julia Farrer, Ian Tyson and Matthew Tyson.

July 1994
JOHN BENTLY
Triptych
Liver and Lights Nos 16, 17, 18

Nov 1994
VANESSA JACKSON /
JAMES LAUGHLIN
A Secret Language
Cast Iron Press

Aug 1994
CHLOE CHEESE
The Eagle
Dinosaur Press /Eagle PH

Dec 1994
SHELF LIFE
including books by Les Bicknell and Andrew Bick.

Jan 1995
MANDY BONNELL
Lamu

Feb 1995
COO GELLER
Cell Book I and II

May 1995
LES BICKNELL
Anticipating Tomorrow
Bookworks in paper and wood.

Carpenter Center for Visual Arts
Mary Halpenny
Harvard University
24 Quincy Street
Cambridge MA 02138 USA
Merely Connect
Tom Phillips
68pp consisting of numerous etchings, digitally composed imagery, cuneiform and Arabic writing and text from Salmon Rushdie's new novel *The Moor's Last Sigh.* Proceeds from the sale of this book will benefit the fund in support of Rushdie's defence against tyranny and the Visiting Artist program at the Carpenter Center.
30.8 x 23 cms. Edition of 12. $3500

Cast Iron Press
1 Morocco Street, London SE1 3HB
A Secret Language
Vanessa Jackson
James Laughlin
Second collaborative work, comprising 12 poems and 6 colour woodcuts, between renowned American avant-garde publisher and poet, now 80, and London painter. Printed letterpress on Somerset, with blocked design on binding.
10 November 1994.
25.4 x 19 cms. Edition of 200.
ISBN 1 898761 01 9. £75.

Circle Press
26 St Lukes Mews,
London W11 1DF
Pages of the Wound
John Berger
This book makes available for the first time a personal selection of John Berger's poems, photographs and drawings (1956-94). Included are 46 poems, nine original photographs, two large photo-litho drawings (32 x 66 cms) and an additional loose signed photograph.
November 1994, London.
33 x 22 cms. In an Edition of 90 signed and numbered.
ISBN 0 901380695. £270.

Coracle
Docking, Norfolk, PE 31 8LQ
Some Notes on Drinking and Driving
Bill Culbert & Simon Cutts
Continues their collaborative project begun with *Some Notes on Writing and Drinking* 1992 (now out of print), which has become a collectors' piece already. This is another vintage piece.
90pp, 7 c, 18 b&w, spiral bound, slipcase. 1994
15 x 10cms. Edition of 200. £20.
The Translated Latrine Inscriptions of the Palazzo Davanzati
Simon Cutts
'..was built around the middle of the 14th century. It contained not only the first sewage systems, but also perhaps the first graffiti in the form of inscriptions on the lavatory walls.' Immaculately researched and presented. (Cutts and Erica VanHorn were based in Florence for a year 1990/1)
36pp, 2-colour photo-copy, hc, 1994
15.5 x 10.3 cms. Edition of 100. £12-50.
The Waterfalls of New Hampshire in Winter
Simon Cutts
A fine bookwork and an exquisite notepad in one. Dedicated to the Van Horns. Two-tone pad of paper, perfect bound, letterpress in plastic binder.
17 x 10.9 cms. Edition of 100. £12-50.
Boy Bell's Book of Envelope Interiors
Erica Van Horn
Exploring the hidden world of envelope interiors. Now in its third edition, but each book is still unique.
28pp, inserts in letterpress cover.
15 x 14.6 cms. Edition 100. 1994. £10.
Special edition: 28pp, casebound, slipcase, edition of 9. £145.
Envelope Interiors Calendar 1995
Erica Van Horn
Now available as a pin-up calendar. The strong design based on squares has minimalist overtones: with its bold lettering it is meant to be used.
24pp, spiral bound. 20 x 20 cms. Edition 100. £12.

Scraps of an Aborted Collaboration
Erica Van Horn
Van Horn is also a born story teller, which is rare among 'serious' bookmakers guarding their reputation. This tragi-comedy is so funny, it just has to be true. The pictures strike the same note.
18pp, offset in three colours, Japanese binding.
11.5 x 19.7 cms. Edition of 200 .1994. £12-50.

Council House Books
34 Five Stiles Road, Marlborough, Wiltshire SN8 4BE.
Hello & Goodbye
Ray Ward
Earth as seen by our own universal guardian / gardener. A humorous thought provoking and highly individual artist's book. An endeavour to explain the black hole phenomenon from a Little England perspective. Hand signed 208 b&w pp, hardback. ISBN 0 9522469 02
Marlborough '94 £13-95 + £3-55 p+p
22 x 31 cms. Edition of 300.

Caine / Petropolis
Atelier de la Cerisaie
31 rue de la Cerisaie, 75004, Paris, France.
Anno Domini 1921
Michael Caine
100 poems of Anna Akhmatova describing the aftermath of the Russian Revolution in 1921. First ever re-edition of this book since 1921, many of the poems were censured in subsequent Russian Editions of Akhmatova's poems. Hand-printed, letterpress.
Paris, Spring 1995.
11.5 x 8.5 cms. Edition of 500. £15.

Meg Campbell
10 Church Lane
Southampton SO2 1SZ
My grandmother, my mother, myself
A bookwork of 6 specimen boxes hung vertically. Pinned into each is an 'insect' representing a generation. The background of each box is made

from layers of history including handwriting. Each insect can be unpinned and read, the wings forming the pages of the book telling the story of that individual.
200 x 30 cms. One-off.

Peter Dover
8 Betts House, Betts Street,
London E1 8HN
Between Two Rivers
32pp, 14 texts, 4 b&w images, full colour cover. Produced by computer, photocopier and printer. Handbound, Meditations, memories and images based on Dover's childhood in and around Liverpool and a one year residency in the city some 20 years later.
1994, Liverpool / London
16 x 17.5 cms. Edition of 200 / 150 paperback £20 / 50 hardback £35.

John Dilnot
11 Harrowby Road
Liverpool L21 1DP
Through the Forest
Take a short walk through the forest. Rubber stamps throughout. Green paper, card cover and spiral binding. 35pp numbered.
1994, Liverpool.
4,5 x 7.5 cms. Edition of 30. £8.
15 Cows
An accordion bound book of 15 cow lino cuts in a numbered edition. Photocopy with a green cover.
1994, Liverpool
6.6 x 6.6 cms. Edition of 500. £5.
Landscape
Open this accordion bound book to reveal a rubber stamped landscape. Green paper with a greener card cover. 15pp, numbered.
1994, Liverpool
5.5 x 5.5 cms. edition of 100. £4.

Dinosaur Press
Eagle Gallery / Eagle P.H
159 Farringdon Road,
London EC1R 3AL
The Eagle
Chloe Cheese
Michael Belben, David Eyre, Emma Hill
An artist's scrapbook of texts, images and recipes. Six colour lithographs and a generous series of b& w images, jostle with handwritten text, creating an affectionate informality in depicting day to day life.
1994, London
22.5 x 20.5 cms. Edition of 100. £125.

Liz Ellis
Britannia Works
Dale Rd, London E3 2NW
Tel 081 555 1672
Some People are too Soft
Concertina book with 8 photographs and text based on fragments of conversation with homeless adults at the East London Centre where Liz runs an art course.
1994, London
17.5 x 12.5 cms. Edition of 10. @ £150.
Also available in Casablanca Magazine, issue no 10, as pull-out book. Send cheque for £2-50 , to Casablanca, 31. Clerkenwell Close, London EC1 OAT

Ken Ferguson
20 Lanercost Park
East Cramlington,
Northumberland NE23 6RM
Helen
Poems by Anne Jones
Poems and images of a teenage daughter. Poems and book cover in relief, 10 intaglio engravings. Poems and text hand-set and book, hand bound. Designed, hand-made, printed and coloured by Ken Ferguson.
1994, Northumberland
17.3 x 12.6 cms. Edition of 33. £75.
Fair View Farm
Collected poems by Anne Jones and Winifred Brassington
26pp of poems and images of a North Wales farm, printed in relief. Poems and text hand-set, images hand-coloured, printed and handbound by Ken Ferguson.
1994, Northumberland
20.3 x 14.8 cms. Edition of 110. £55.
My Friend John
Poems by John Payne
10 poems by John Payne. 10 intaglio engravings. Letterpress. Hand- set etc. by Ken Ferguson.
1994, Northumberland
17.3 x 12.6 cms. Edition of 60. £65.

Far North Ceramics
Durness,
Sutherland N27 4PT
Fragment of Land
A collaboration between Lotte Glob and Marshall Anderson.
This series of unique bookworks explores the book as object. Each work is concerned with placing the natural landscape within the cultural landscape and is made by vitrifying rocks, sediments in combination with other materials.
January 1995, Durness
46 x 30 cms. Series 20. £750.

Gefn Press
5. Elmwood Road, Herne Hill
London SE24 9NU
Crevice / Map
Susan Johanknecht and Jennifer Newson
Crevice / Map deals with relationships between internal and external landscapes using imagery from ordnance survey maps and body organs. The text uses language of geology and of a journey to describe the vulnerability of bodies and landscapes.
1994, London
35 x 18 cms. Edition of 80. £60 during duration of Hardware Gallery exhibition (24 April - 15 May 1995) £80 thereafter.

ANDI McGARRY
SUN, MOON +STARS PRESS
DONKEY MEADOWS
WEXFORD
EIRE

Trial of the Unseasoned Mariner
A set of 7 Boat Books
Colour photos, ink drawings throughout, entirely handmade
11 x 15 cm . 20-34 p.p.
£10 each -- £60 for set
Nod
16 p.p - -£10
The Great Great Great Irresistible
14p.p -- £10
Sea Dive
14 p.p. -- £7-50
Fishy Ones
10 p.p. -- £10
(All inc p+p)

THE NO INSTITUTE
JURGEN O. OLBRICH
BODELSCHWINGHSTR . 17 ,
34119 KASSEL GERMANY

No News
Object / magazine published since 1986. Each issue made in collaboration with one artist and varies in shape, size, layout and edition number. All copies are normally originals

No.26 **Stewardess-box,** with Figueiredo
80 copies / DM 100
No.27 **Culture bag,** with The Nomads
60 copies / sold out.
No.28 **Life style,** with R. Nissle
100 copies / DM 80
No.29 **Copper Institute,** U. Idelberger
75 copies / DM 80
No.30 **Box Sculpture** with AY-O
75 copies / DM140
No.31 **Ingredients** with F. Zobel
99 copies / DM 75

NATALIE d'ARBELOFF
6. LADY SOMERSET ROAD
LONDON NW5 1UT
(0171) 267 1719

THE CREATION
from the Book of Enoch
NdA Press 1992
43.5 x 34 cms. Edition of 12
(Two in colour, similar but not identical, 10 in black). 39 intaglio plates plus relief-printed wrapper and box cover plate. Text printed intaglio from zinco relief blocks on BFK 180gsm (black version), Arches 250 gsm (colour version) Copies 1-9: 20 double leaves in wrapper held in felt-lined solander box with one of the etchings printed relief on the cover. Price: £2200
The first eight days of creation as told by the Creator to his 'scribe' Enoch, in a little-known text, mysterious and powerful, written around the beginning of the Christian era, in Egypt in Greek, later translated into Slavonic, and discovered only in the 1920s in manuscripts found in Russia and Serbia. The English translator is not known. NdA selected only some extracts for her illustrated version of this extraordinary book.

TI PARKS
ARTISTS BOOKS
9 THE ORCHARD
LONDON SE3 OQS
(0181) 852 7246

'SETS OF SAM' AND 'OOM WITH BA' UNIQUE 'ROAD PRINT' BOOKS (£200); **'RADIO' AND 'ENCORE ANGORA'** IN EDITIONS OF 4, TEXTLESS BOOKS OF SOMETIMES EROTIC IMAGES FROM THE ARTIST'S MAJOR WORK **'10,000 COLLAGES' (£200)** **'NET'** IN AN EDITION OF 4, A TWELVE-PART 'BOOK SCULPTURE' (£200); **'4B CROSSDRAWINGS (FOLDED IN)' & '4B CROSSDRAWINGS (FOLDED OUT)'** IN EDITIONS OF 20, WALLETS OF ORIGINAL PENCIL DRAWINGS (£40). THIS IS ONLY A SMALL SELECTION OF RECENT TITLES. A FULL LIST AND MORE INFORMATION, IS AVAILABLE FROM TI PARKS.

HARDWARE GALLERY

162 Archway Road,
London N6 5BB
Curated by Deidre Kelly
0181 341 6415
Fax 0181 348 0561
Gallery Opening hours are
Tues - Sat 2-6pm
or by appointment.

Hardware Gallery was established in 1986 in North London. Since that time, Hardware has gained a reputation as a specialist in Contemporary Printmaking and Artists' Books. A permanent feature of Hardware, is its collection of British artists' publications, which show great variety in method and application to bookworks. 1995 programme includes new works by Susan Johanknecht, Simon Lewandowski, Les Colman amongst others.

May 1994
LES BICKNELL
'The Ordinary made Extraordinary'

'Looking at Words: Reading Pictures'.
Artist publications from the U.K. Curated by
David Blamey.

June 1994
'Unfolding'
featuring
bookworks by
MARION BRUCE

Oct 1994
IAN TYSON
Tetrad Press
'Some of a kind'

March 1995
Concourse Gallery Barbican
Focus on Artists' Books at the Contemporary Print Show.

May 1995
Book Installation by
SUSAN JOHANKNECHT AND JENIFER NEWSON
'Crevice - Map'

Hardware Gallery
162 Archway Road,
Highgate,
London N6 5BB
Looking at Words: Reading Pictures
Edited by David Blamey
The publication which accompanied the 'Looking at words: Reading pictures' exhibition. 32 artists including: Art in Ruins, Henry Bond, Daniel Jubb & Tim Lewis. Available in three colour combinations.
May 1994, London
21.5 x 16 cms. Unlimited edition
ISBN 0 949004 11 1 £1-50.
The Story of Guaranà
Deidre Kelly
The mythical story of the Guaranà fruit from Brazil. Screenprinted leporello with some collage. Produced while in Brazil on a British Council exchange project at 'Museu da Gravura' Curitiba.
April 1995
15.2 x 10.67 cms (when fully open).
Edition 10. £45.
Homage to Calvino
Deidre Kelly
8 Pages, collage and watercolour images after the writing of Italo Calvino.
1994, London
15.2 x 14.5 cms. Unique. £65.

Andrea Hill
117 Vanbrugh Hill
Greenwich, London SE10
The Ship and the Shore
Embedded miniature book in plaster with shells, starfish and found ceramic pieces salvaged from the tide of the Thames River. Encased inside wooden box with dancing figure. hand painted with inks and dyes.
Summer 1994
19 x 17 cms (Book inside box measurement) One-off. £150.

Christopher J.A. Houlding
115 Tottenham Lane, Crouch End, Hornsey London N8 9BW
Glass Dome Book
An admission register book with black hard covers & maroon cloth spine. 128 pages are printed to maintain a register. A glass dome (such as might be used to cover cheese) has been attached to front cover.
1994, London
14.5" x 11.25" x 8" (deep). One-off.
Medical Instrument Book
An admission register book with black hard cover and maroon cloth spine. 128 pages are printed as for register. A pair of forceps have been inserted into the top of the book and syringe & part of a stethoscope attached to the front cover of book.
1994, London
14.25" x 11.25". One-off.
The Master's Masterpieces
Dark green cloth covered book inscribed with gold lettering. Book originally published by William Heinemann in 1903. 200 plus golden pages filled with b&w photographic images of dead naked men. Book has been produced to commemorate 80th anniversary of start of Great War.
17" x 12.5". One-off
None of Houlding's books are for sale.

Imprints
Le Vieux Village
26400 Piégros La Clastre
France.
A Wish
Matthew Tyson & Robert Creely
7 Painted pages with letterpress text
1994
19 x 37 cms. Edition of 5. FF 3500.
Vacant Possessions
23 Potato and card printed pages with letterpress text. 1994.
13 x 19 cms. Edition of 12. FF 1500
Breath
Mark Hudson
Cardboard prints and letterpress text. 1994
16 x 26 cms. Edition of 6. FF 650
Books are Important
Matthew Tyson and Les Bicknell
Catalogue comprising 24 screen-printed and offset pages, stitched with folded grey paper cover. Essay in French and English by Rowan Watson.
Pub. Maison du Livre d'Artiste Contemporain.
40 x 17.5 cms. FF60.

Jolly Roger Books
The Workshop, Bath Old Road
Radstock, Bath Avon BA3 3HD
The Incredible Bike Man
Jim Roberts
A fictional narrative, inspired by a passion for cycling, combining images and text in a series of 24 reproduced relief prints. Printed in b&w on recycled paper .
Aug 1994, Bath, Avon.21 x 15 cms.
Edition of 200. £3-50 in p&p.

Kamikaze Press
149 Archway Road, London N6 5BL
Bust
Simon Redington
A concertina style book using headline type to relate a collection of words and phrases used on the street. For example 'Blue Grenade', 'Bull-ring box-up' + 'Jack' are toxic substances which provide oblivion. The type is printed several times to form a chaotic barrage of language.
1994, London
9" x 7" Edition of 30. £150.
Hostel
Simon Redington
A portfolio of 15 prints about homelessness in London. Containing a mixture of type and image suggesting the tragic ironies inherent in the institution of the hostel and the Dickensian existence it continues to represent. The sign above a shouting man's head states 'Commit no Nuisance'
1994, London
25" x 19". Edition of 30. £1500.

Tony Kemplen
29 Meadowhead
Sheffied S8 7UA
A long tour of Paris by Metro
56 screen-printed pages giving directions for a journey on the Paris metro system, in homage to Gustav Eiffel. Sewn in seven signatures onto cords, which in turn are formed into the shapes of the metro lines beneath the leather covering.
1995, Sheffield
29 x 51 cms. Edition 4 (Bindings all different) £400.
A Time to..
A visual poem, 144 pages screen-printed in four colours with added inkjet text. The hours of the day pass by against a background rhythm of dandelions and clocks.
1995, Sheffied.
15 x 15 cms. Edition of 6. £100.
SK329858 1944-1994
introduction by Sharon Kivland
The book takes the form of a pack of playing cards, which tells in a variety of ways the story of the wartime crash of an American bomber in a Sheffield Park. Boxed with a 16 page booklet.
1994, Sheffield.
8.8 x 5.8 cms. Edition of 100. (In conjunction with Sheffield Hallam University) £10.

Liver and Lights
John Bently
120 Goodrich Rd
London SE21 OER
Liver and Lights no. 14.
A book of Discarded things.
A book made entirely of rubbish collected in the vicinity of the artist's studio in Deptford. A second edition of 60 copies. £45.
Liver and Lights no. 16-18
Triptych July 1994
3 linked bookworks in an edition of 60. £90 "In response to the seemingly eternal chain of refugees criss-crossing Europe, these books form an interminable road on which history perpetually repeats itself."
Liver and Lights no. 19-20
Deptfordia. July 1995
Edition of 60. A series of songs about Deptford, a mythic place, a book-like vessel in which a multitude of characters related to Deptford, the artist's beloved place, sail... £60.

Jean Lodge
Ruskin School of Art,
74 High Street, Oxford OX1 4BG
Expectations
Jean Lodge
Alan Chatham de Bolivar
28 pages plus cover, 14 woodcuts, manuscript text, stitching and collage, in cloth covered box. Part of a series "A mano"
1994, Paris
8.5" x 8.7". Edition of 9. £500
Galerie André Biren
11 Rue Cassette,
75006 Paris, France
Un Autre, Another
Jean Lodge
Geneviève Bouchiat and Jean Lodge
Bilingual, text in French and English, 22 double pages, 4 colour etchings, typography by Pierre Mréjen.
1994, Paris
11.5" x 8.8". Edition of 300. £180.
Editions du Rouleau Libre
10 Rue de Vaucouleurs
75011 Paris, France.

M Press,
Mikey Georgeson
Rosetta Studios
36 Great Russell Street,
London WC1 3PP
From the Heart
Superimposes hearts onto everday words and objects, to reveal a tale of true love and betrayal. Originally conceived as an elaborate Valentine's card, this book was a commercial flop due to the down-beat, bitter-sweet ending.
8 pages, two colour p/c. £1-50
Gratuitous
M Press's latest offering, conceived as a periodical, it is already into its second issue. The images are purely arbitrary, somehow having floated to the surface, whereupon they ask politely to be rendered. Comes with free badge. *Bypass* magazine saw it as a "tirade against narrative"
28pages, three colour p/c. £3.

Magpie Press,
1. Hermitage Cottage, Clamp Hill,
Stanmore, Middx HA7 3JW
Journey: From Eritrea to Edgware
Text by Jediyon Ghebremariam
A collaborative book project in English and Arabic.
Designed by Frontiers Design (Nottingham)
24pp - 12 colour and a full colour cover.
January 1995, London
21 x 15 cms. Edition of 500.
ISBN 0 9523880 57. £3-50 plus 50p p&p
Following
Tanya Peixoto and Linda Jenkins
Extracts from poems and images
Hand-made books, colour photocopies
December 1994
15 x 21 cms. Edition of 24. £10.

B. Marjoram
B.M Box 8614
London WC1N 3XX
Beautiful, Elegant, Tasteful
The text of this book offers a series of questions and asides which have a playful childlike quality, sometimes obscure, sometimes wanting to shock. 1994. 12pp bound in blue leather. 13" x 19.75".
Edition of 15. £120.
My Hostility
A book sculpture with a cover in silver leaf; contains a strip of images and captions centred on English clichés, games, themes, inspired by magazine illustration from the 30s and 40s and ideas of correctness, manners, and propriety which have at their core self-restraint and denial.
Jan 1995. 5 pp, 25" x 8.5".
Edition of 15. £80

DAVID BARTON
45 WELLMEADOW ROAD
HITHER GREEN
LONDON SE13 6SY

"My books contain events from a continuing involvement in the creation of my life through drawing. Throwing myself away into the empty embrace of the page, I search within it for transformation and renewal. By accepting and exploring the images which appear, I achieve glimpses of the thing I am.".
68 Books published to date plus 7 in collaboration with poets. All books are signed and numbered.

STUNT, Ed. 500. 32pp
STUNT 2, Ed. 500. 32pp
40 STRIPS , Ed. 60. 20pp
24 STRIPS , Ed.50. 12pp
28 STRIPS , Ed. 50. 14 pp
50 STRIPS , Ed. 50. 25 pp
38 STRIPS , Ed. 50. 19pp
32 STRIPS , Ed. 50. 16pp
STAMPING , Ed. 500. 336 pp
30 STRIPS , Ed. 50. 15pp
US M.Howlett, Ed. 50. 15pp

THE INCREDIBLE BIKE MAN by JIM ROBERTS

A5 offset chapbook 32 pages printed on recycled paper in a signed edition of 200. Inspired by the artist's passion for cycling, this visual narrative incorporates images and text, in a series of reproduced relief prints.
Available at **£3-50** each including Postage and packing from:

JIM ROBERTS
The Workshop, Bath Old Road
Radstock, Bath, Avon BA3 3HD

CREVICE / MAP

by Susan Johanknecht and Jenifer Newson

Crevice / Map deals with relationships between internal and external landscapes using imagery from ordnance survey maps and body organs. The text uses language of geology and of a journey, to describe the vulnerability of bodies and landscapes.
35 x 18 cm (145 x 61 cm unfolded)
Edition of 80. £80 each. 1994.

Published by **GEFN PRESS,**
5 ELMWOOD ROAD, HERNE HILL
LONDON SE24 9NU.

BRAIN CELL

PLEASE SEND ME YOUR STAMP DESIGN, RUBBER STAMP OR 150 STICKERS OR SEALS. I WILL PRINT OR PASTE THESE MATERIALS ONTO A3 PAPER, CREATING 150 SHEETS. I WILL THEN SEND A SHEET BACK WITH A LIST OF ADDRESSES TO EACH PARTICIPANT. I WILL PUBLISH AT INTERVALS OF 8 TO 10 DAYS AND THE ISSUE AT THAT TIME WILL INCLUDE 60 PEOPLE OR SO. BRAIN CELL IS ALWAYS SEEKING A CHANGE, DOES NOT INTEND TO SETTLE AND CARE FOR ITS EXTENSION OF THE NETWORK SO DON'T MAIL A LUMP OF STUFF FOR SEVERAL ISSUES. PLEASE SEND THEM TO ME FOR ONE ISSUE AT A TIME.

SEND TO: RYOSUKE COHEN,
3 - 76 -I -A -613 YAGUMOKITACHO,
MORIGUCHI CITY,
OSAKA 570, JAPAN.

WORKFORTHEEYETODO

51 Hanbury Street
London E1 5JP
Curated by Maggie Smith
0171 426 0579
Gallery Opening hours are
Thurs - Sat
11am-6pm
and by appointment

Workfortheeyetodo has been open since 1993. They see their function as both providing information and making available a wide selection of current and empirical books. At the same time they run a continuous programme of displays and arrangements to extend and complement critical concerns of the book as a medium for artists, with exhibitions by Colin Sackett, Johnathon Williams and Thomas Meyer, Ian Hamilton Finlay, and Henry Miller.

for it not to be worth the paper it is printed on
it has to be printed for it not to be worth the
paper it is printed on it has to be printed for
it not to be worth the paper it is printed on it
has to be printed for it not to be worth the pap
er it is printed on it has to be printed for it
not to be worth the paper it is printed on it
has to be printed for it not to be worth the paper
it is printed on it has to be printed for it not
to be worth the paper it is printed on it has to
be printed for it not to be worth the paper it
is printed on it has to be printed for it not to
be worth the paper it is printed on it has to be
printed for it not to be worth the paper it is
printed on it has to be printed for it not to be
worth the paper it is printed on it has to be printed

Les Coleman
1994

April 1994
'For the Voice: Sounds and Silence in Artists' Books'
Curated by Dr Stephen Bury
Librarian, Chelsea College of Art and Design. Catalogue.

June 1994
'Ten Years of Atlas Press'
Emissions of the Anti-Tradition.
Catalogue.

Aug 1994
'Some Implications of Poetry'
Catalogue.

Nov 1994
'Interaction and Overlap'
The Little Magazines, Poetry Store, and Alternative Press Collection.
University of London College Library. Catalogue.

Oct 1994
'Some Notes on Drinking and Driving'
SIMON CUTTS & BILL CULBERT

Jan 1995
A Selection of New Books from Europe.

Morning Star Publications
Alec Finlay
17 Gladstone Terrace
Edinburgh EH9 1LS
The Morning Star Folios: Fifth Series 1994
Collaborations between artists and poets from around the world.
Folio 5/1 *Birds and Flowers,* poems by Michael Longley, illustrated with a watermark
Folio5/2 *Palabrar mas, / Wurdwappin schaw,* visual poems by Cecilia Vicuna, translated into Scots by Edwin Morgan.
Folio 5/3 *A Language I do not know,* poems by Jaan Kaplinski drawings by Edda Renouf
Folio 5/4 *Morning and Evening,* poems translated from the Japanese by Sam Hamill, drawing by David Connearn
Editions of 300 including 26 copies signed and lettered by the contributors.
Each folio comes in a C5 printed series envelope.
Subscriptions (four issues) £20. Individual issues £6. Signed issues £12
Under the Moon Series: First Series 1994
Issue 1/1 *A Semblage of Steerage,* rudder and lever poems by Ian Stephen, drawing by Will Maclean
Issue 1/2 *Impending Navigation Bright,* Poems about the shipping forecast with a drawing by Margot Sandeman.
Issue 1/3 *The Obscure Voice,* translations from Italian poetry by Hamish Henderson, illustrated by Ian McKeever
Issue 1/4 *In the Blue Mountain Evening,* poems by Friederike Mayrocker.
Editions of 300 copies, including 26 copies, signed and lettered by the contributors. Folded concertina with A5 printed manilla cover.
Subscriptions (four issues) £18. Individual issues £5. Signed issues £10. All titles are edited and designed by Alec Finlay.

Andrew Norris
25 Wimbourne Court, Wimbourne Street, London N1 7HB
Herd
12 pages of ink drawings of Fresian cows. A hand-drawn frieze on Arches paper. Cover bound in simulated cow hide, boxed, also wrapped in same.
1994, London
10 x 13 cms. Edition of 20. £27.
Wheatsheaf
One of a continuing series of small books, average 32 colour photocopied images, collected from food packaging which uses a poetic rural idyll as a means of marketing the product.
1994, London
6 x 8 cms. Not editioned.
£3-50 inc p&p.

The Oleander Press
17 Stansgate Avenue,
Cambridge CB2 2QZ
Contemporary Designer Bookbinders: An Illustrated Directory
Edited by Philip Ward
The first ever illustrated worldwide survey,, showing artist bookbinders (not conservers or restorers) and their masterpieces, with biographical profiles and their aesthetic statements.
August 1995
21 x 17.5 cms
ISBN 0 906672 36 8. £30.

Mark Pawson
P.O.Box 664, London E3 4QR
All My Rubber Stamps
Every rubber stamp Pawson's used since childhood - over 250 in total, stamped by hand in classified sections and full colour section. Price includes a free product upgrade. hardback copies are sewn and bound in black sheet rubber, each copy has uniquely stamped end papers. 80pp
1994/95, London
21 x 15 cms. Edition of 250. £15. Hardback (30) - £45

Penny Press
1 Morocco Street
London SE1 3HB
Co Mayo
Norman Ackroyd
Twelve aquatint etchings resulting from a visit to Ballycastle in North Mayo as a guest of the Ballingham Arts Foundation. Housed with a map and title pages in a Solander box.
December 1994
35 x 28 cms. Edition of 35. £900.

Ian Pyper
8 Roedean Court
Roedean Rd, Brighton
East Sussex BN2 5RT
Hospital of the Angels
Ten images of Angels inspired by the 'Hospital of the Angels' in Mexico City which is currently carrying out surgery on foetuses in the womb.
March 1995
A4. Edition of 25. £8-25

Ti Parks
9 The Orchard
London SE3 OQS
Corner to Corner
Eight colour, corner to corner, pencil cross drawings, using both sides of 200-220 sheets. Pencils chosen at random so that each page and each book is different. Bound with no cover. Title sheet, signed and numbered, near the middle.
30 x 21 cms. Edition of 10. £350
4B Crossdrawings (Folded in)
Four folded sheets contained in a clear plastic pocket with title card signed and numbered. each sheet has a corner to corner pencil cross drawing folded in, and is designed to be opened out but not flattened.
1994, London
16.5 x 11.5 cms. Edition of 20. £40.
4B Crossdrawings (folded out)
The same as *4B Crossdrawings (Folded in),* but the four drawings are folded out.
1994, London
16.5 x 11.5 cms. Edition of 20. £40

Real Art
Gilesway, How Mill, Carlisle
Cumbria CA4 9JT
Real Art
Edited by Malcolm Gibson
Invited artists regularly include Malcolm Gibson, Rachel Gibson, Andrew Law, Elspeth Law, James Hall. Published twice yearly, this magazine is devoted entirely to visuals. Up to ten artists contribute work: screenprints, lithographic, blockprints, as well as collaged and handmade materials.
A4. Edition of 300. £10.

R.M.G. Publishing
406 St Johns Street
London EC1
Failure
Andrew Bick
A collaboration between Andrew Bick and RMG consisting of text, lithographs, monotypes and drawings on the subject of 'failure'
"I'm failing, he thought. But this failure was not disagreeable to him, on the contrary it gave him a special feeling of happiness such as he had never had before." Leo Tolstoy.
1994, Den Haag. March/April 1995, London.
25 x 32.5 cms. Edition of 100. £50.
Adventures in the West
Christopher Twig
'A collection of poems written between 1991 - 1993. The first 200 copies in the edition have been signed, stamped and numbered with original drawings on each cover.
June 1994, Eagle Gallery, London
15.5 x 11.5 cms. Edition of 1000
ISBN 899245 006, signed edition (200)
ISBN 899245 057. £5.

Red Hen Press
Shirley Jones
Byddwn Uchaf
Llanhamlach, Brecon,
Powys LD3 7SU
Pick Me a Bunch of Roses
A selection of poems and translations by Shirley Jones with some of her images reproduced from earlier works.
9.5" x 6.5". Edition of 500. £7-50.

Childe Roland
One Glan Aber, Hill Street,
Llangollen, Clwyd
Wales. LL20 8EW
I-dot
Hard cover recycled remainder book with an aperture in the front cover which lets in light in order to dot a candle-shaped letter 'i' on the fly-leaf inside.This book is to be launched in August at the Glyn Vivian Gallery in Swansea as part of a bookwork show *Beyond Reading* organised by Les Bicknell.
1995, Llangollen
17.7 x 16.5 cms. Edition of 75. £7.

Deb Rindl
22 Cardigan Road
Bow, London E3 5HU
The 4 Seasons
Set of four books, one for each season, expressing the essence of the season through folding and letterpress printing.
1994, London
10.2 x 35 cms. Edition of 4 sets, *signed and numbered. £100 per set.
Heard the Horse
Concertina-style book in black slip-case, containing poem by Tom Hooper, concerning destruction of animal species. Printed letterpress and hand folded and finished. Presented in a black box
1995, London
14.2 x 14.2 cms. Edition of 25, signed and numbered. £30.
Parallel Lives
Two-sided book concerning the nature of our day-time and night-time lives; each one only very faintly aware of the other. Printed letterpress and hand folded and finished. Presented in a black box.
1995, London
17.5 x 17.5 cms. Edition of 20. £35.

Sun, Moon and Stars,
Donkey Meadow,
Kilmore Quay,
Wexford, Eire.
Notes on Time
Andi McGarry
Handmade, Indian Ink, Bristol Board, "hop step and jump through Time's joyful contract."
1995, Wexford.
4" x 3.5". Edition not more than 30. £15.

Roger Towndrow
94 Marlborough Road
Falmouth
Cornwall
TR11 3LR
a place apart
Through imagery the book expresses an emotive response to place, distance and desire using the artist's studio and memories of travel as reference. 20 pages. 8 single colour block printed images on Arches Text Wove. Japanese flyleaf. cardboard covers. handbound, letterpress text.
1994, Cornwall.
22 x 35 cms. Edition of 25. £75.

Jeroen van Westen
Reelaan 25
7522LS
Enschede
The Netherlands
(053) 351098
Orientation
A set of four maps each glued in a linen cloth hard cover. "The maps show four constellations of references in words as made in my studio. Owning a set is an invitation to (re) create a personal interpretation of orientation."
*1994, Enschede
22 x 13.7 cms. Edition of 14. £300.
Itinerario
A box in which there are two

descriptions of 'itinerarios' on geese. One a spirally folded two-sided map. The second is an audio-cassette that can guide a short (spirally) walk when listened to with earphones.
1994, Enschede
9.3 x 13.3 cms. Edition of 14 odd nos Dutch, Even nos English. £300.
Atlas
An Atlas in which van Westen explores the constellations as in *Orientation.* The maps are monoprints, with transparent overlays of the constellations. Each Atlas presented as a semi-circle (106cms across) reading standard, the book resting against two footprints in the standard.
1994, Enschede
39 x 29.3 cms. Edition of 5. £3000

Weproductions
Deuchar Mill,
Yarrow
Selkirkshire
Scotland TD7 5LA
Water on the Border
Helen Douglas & Telfer Stokes
Establishes connections between Scotland and China by combining drawings of trees, plants and insects made by primary school children with photographic images of water taken from the Yarrow Water, Selkirk and West Lake, Hangzhou. Interspersed with 6 Chinese poems transcribed from Chinese into Scots by Brian Holton this book draws inspiration from T'ai Chi and Su Causeway, Hangzhou to create a continuum and evoke the confluence between water and the water's edge.
1994, Scotland
19 x 14.5 cms. Edition of 600.
ISBN 0 9523284 96. £15.

Welfare State International
The Ellers,
Ulverston
Cumbria, LA12 OAA
Capsule
Tanya Peixoto
The words and voices of ex-pupils (in their eighties) of the National School, Ulverston, now the headquarters of WSI. Collaged together with images by John Fox, Peixoto and primary school children. Layers of images and text evoke the process of 'remembering'
19 x 19 cms. Edition of 300. Local price £2-50, National £7-50.

Writers Forum
89a Petherton Road,
London N5 2QT
Light
Bob Cobbing, Pierre Garnier, Arrigo Lora-Totino, Bill Keith and Hiroshi Tanabu
Five visual poems, five poets, published simultaneously in five countries.
English edition, March 1994, London.
25.7 x 18.25 cms. Edition of 100.
ISBN 0 86162 5382. £1-25.
Partial Additives
Bob Cobbing
Paul Dutton
The witty addition of a letter to a word to change one 'substance' into another, graphically illustrated.
21 x 14.85 cms. Edition of 200
ISBN 0 86162 551 X. £2.
Domestic Ambient Moise (an ongoing sequence)
Bob Cobbing, Lawrence Upton
Lawrence Upton, Bob Cobbing
Visual text by Upton or Cobbing with six variations on this theme by Cobbing or Upton (nine in series with same title so far published)
Nov 1994-Feb 1995, London
25.7 x 18.25 cms. Edition of 100.
ISBN 086162 5587, 562 5, 5633,5688, 5676, 5684, 569 2, 5714, 5722. £1 each.